DEDICATION

This book is dedicated to all the ambitious and organized people out there who love to keep the vintage art of Ham Radio or Amateur Radio alive.

Your are my inspiration for producing books and I'm honored to be a part of keeping all your Ham Radio notes all in one place.

This journal notebook will help you record your details for using your Amateur or Ham Radio.

Thoughtfully put together with these sections to record: Date/ Year, Time UTC Start & Finish, Frequency MHz, Mode, Power, Station, Report Sent & Received, Signal Sent & Received, Remarks/ Notes.

HOW TO USE THIS BOOK

The purpose of this book is to keep all of your lists and notes all in one place. It will help keep you organized.

This Ham Radio Journal will allow you to accurately document .

Here are examples of the prompts for you to fill in and write about yourself in this book:

1. Date / Year

2. Time UTC Start & Finish

3. Frequency MHz

4. Mode

5. Power

6. Station

7. Report Sent & Received

8. Signal Sent & Received

9. Remarks / Notes - For writing any important details such as map, equipment used, experience, test, use for work or vintage hobby, etc.

AMATEUR RADIO STATION LOG

Date	Time UTC		Frequency (MHz)	Mode	Power	Station	Report		Signal	
	Start	Finish					Sent	Received	Sent	Received

Remarks

Date	Time UTC		Frequency (MHz)	Mode	Power	Station	Report		Signal	
	Start	Finish					Sent	Received	Sent	Received

Remarks

Date	Time UTC		Frequency (MHz)	Mode	Power	Station	Report		Signal	
	Start	Finish					Sent	Received	Sent	Received

Remarks

Date	Time UTC		Frequency (MHz)	Mode	Power	Station	Report		Signal	
	Start	Finish					Sent	Received	Sent	Received

Remarks

Date	Time UTC		Frequency (MHz)	Mode	Power	Station	Report		Signal	
	Start	Finish					Sent	Received	Sent	Received

Remarks

Date	Time UTC		Frequency (MHz)	Mode	Power	Station	Report		Signal	
	Start	Finish					Sent	Received	Sent	Received

Remarks

Date	Time UTC		Frequency (MHz)	Mode	Power	Station	Report		Signal	
	Start	Finish					Sent	Received	Sent	Received

Remarks

AMATEUR RADIO STATION LOG

Date	Time UTC Start	Finish	Frequency (MHz)	Mode	Power	Station	Report Sent	Received	Signal Sent	Received

Remarks

Date	Time UTC Start	Finish	Frequency (MHz)	Mode	Power	Station	Report Sent	Received	Signal Sent	Received

Remarks

Date	Time UTC Start	Finish	Frequency (MHz)	Mode	Power	Station	Report Sent	Received	Signal Sent	Received

Remarks

Date	Time UTC Start	Finish	Frequency (MHz)	Mode	Power	Station	Report Sent	Received	Signal Sent	Received

Remarks

Date	Time UTC Start	Finish	Frequency (MHz)	Mode	Power	Station	Report Sent	Received	Signal Sent	Received

Remarks

Date	Time UTC Start	Finish	Frequency (MHz)	Mode	Power	Station	Report Sent	Received	Signal Sent	Received

Remarks

Date	Time UTC Start	Finish	Frequency (MHz)	Mode	Power	Station	Report Sent	Received	Signal Sent	Received

Remarks

AMATEUR RADIO STATION LOG

Date	Time UTC		Frequency (MHz)	Mode	Power	Station	Report		Signal	
	Start	Finish					Sent	Received	Sent	Received

Remarks

Date	Time UTC		Frequency (MHz)	Mode	Power	Station	Report		Signal	
	Start	Finish					Sent	Received	Sent	Received

Remarks

Date	Time UTC		Frequency (MHz)	Mode	Power	Station	Report		Signal	
	Start	Finish					Sent	Received	Sent	Received

Remarks

Date	Time UTC		Frequency (MHz)	Mode	Power	Station	Report		Signal	
	Start	Finish					Sent	Received	Sent	Received

Remarks

Date	Time UTC		Frequency (MHz)	Mode	Power	Station	Report		Signal	
	Start	Finish					Sent	Received	Sent	Received

Remarks

Date	Time UTC		Frequency (MHz)	Mode	Power	Station	Report		Signal	
	Start	Finish					Sent	Received	Sent	Received

Remarks

Date	Time UTC		Frequency (MHz)	Mode	Power	Station	Report		Signal	
	Start	Finish					Sent	Received	Sent	Received

Remarks

AMATEUR RADIO STATION LOG

Date	Time UTC Start	Finish	Frequency (MHz)	Mode	Power	Station	Report Sent	Received	Signal Sent	Received

Remarks

Date	Time UTC Start	Finish	Frequency (MHz)	Mode	Power	Station	Report Sent	Received	Signal Sent	Received

Remarks

Date	Time UTC Start	Finish	Frequency (MHz)	Mode	Power	Station	Report Sent	Received	Signal Sent	Received

Remarks

Date	Time UTC Start	Finish	Frequency (MHz)	Mode	Power	Station	Report Sent	Received	Signal Sent	Received

Remarks

Date	Time UTC Start	Finish	Frequency (MHz)	Mode	Power	Station	Report Sent	Received	Signal Sent	Received

Remarks

Date	Time UTC Start	Finish	Frequency (MHz)	Mode	Power	Station	Report Sent	Received	Signal Sent	Received

Remarks

Date	Time UTC Start	Finish	Frequency (MHz)	Mode	Power	Station	Report Sent	Received	Signal Sent	Received

Remarks

AMATEUR RADIO STATION LOG

Date	Time UTC		Frequency (MHz)	Mode	Power	Station	Report		Signal	
	Start	Finish					Sent	Received	Sent	Received

Remarks

Date	Time UTC		Frequency (MHz)	Mode	Power	Station	Report		Signal	
	Start	Finish					Sent	Received	Sent	Received

Remarks

Date	Time UTC		Frequency (MHz)	Mode	Power	Station	Report		Signal	
	Start	Finish					Sent	Received	Sent	Received

Remarks

Date	Time UTC		Frequency (MHz)	Mode	Power	Station	Report		Signal	
	Start	Finish					Sent	Received	Sent	Received

Remarks

Date	Time UTC		Frequency (MHz)	Mode	Power	Station	Report		Signal	
	Start	Finish					Sent	Received	Sent	Received

Remarks

Date	Time UTC		Frequency (MHz)	Mode	Power	Station	Report		Signal	
	Start	Finish					Sent	Received	Sent	Received

Remarks

Date	Time UTC		Frequency (MHz)	Mode	Power	Station	Report		Signal	
	Start	Finish					Sent	Received	Sent	Received

Remarks

AMATEUR RADIO STATION LOG

Date	Time UTC Start	Finish	Frequency (MHz)	Mode	Power	Station	Report Sent	Received	Signal Sent	Received

Remarks

Date	Time UTC Start	Finish	Frequency (MHz)	Mode	Power	Station	Report Sent	Received	Signal Sent	Received

Remarks

Date	Time UTC Start	Finish	Frequency (MHz)	Mode	Power	Station	Report Sent	Received	Signal Sent	Received

Remarks

Date	Time UTC Start	Finish	Frequency (MHz)	Mode	Power	Station	Report Sent	Received	Signal Sent	Received

Remarks

Date	Time UTC Start	Finish	Frequency (MHz)	Mode	Power	Station	Report Sent	Received	Signal Sent	Received

Remarks

Date	Time UTC Start	Finish	Frequency (MHz)	Mode	Power	Station	Report Sent	Received	Signal Sent	Received

Remarks

Date	Time UTC Start	Finish	Frequency (MHz)	Mode	Power	Station	Report Sent	Received	Signal Sent	Received

Remarks

AMATEUR RADIO STATION LOG

Date	Time UTC Start	Finish	Frequency (MHz)	Mode	Power	Station	Report Sent	Received	Signal Sent	Received
Remarks										

Date	Time UTC Start	Finish	Frequency (MHz)	Mode	Power	Station	Report Sent	Received	Signal Sent	Received
Remarks										

Date	Time UTC Start	Finish	Frequency (MHz)	Mode	Power	Station	Report Sent	Received	Signal Sent	Received
Remarks										

Date	Time UTC Start	Finish	Frequency (MHz)	Mode	Power	Station	Report Sent	Received	Signal Sent	Received
Remarks										

Date	Time UTC Start	Finish	Frequency (MHz)	Mode	Power	Station	Report Sent	Received	Signal Sent	Received
Remarks										

Date	Time UTC Start	Finish	Frequency (MHz)	Mode	Power	Station	Report Sent	Received	Signal Sent	Received
Remarks										

Date	Time UTC Start	Finish	Frequency (MHz)	Mode	Power	Station	Report Sent	Received	Signal Sent	Received
Remarks										

AMATEUR RADIO STATION LOG

Date	Time UTC Start	Finish	Frequency (MHz)	Mode	Power	Station	Report Sent	Received	Signal Sent	Received

Remarks

Date	Time UTC Start	Finish	Frequency (MHz)	Mode	Power	Station	Report Sent	Received	Signal Sent	Received

Remarks

Date	Time UTC Start	Finish	Frequency (MHz)	Mode	Power	Station	Report Sent	Received	Signal Sent	Received

Remarks

Date	Time UTC Start	Finish	Frequency (MHz)	Mode	Power	Station	Report Sent	Received	Signal Sent	Received

Remarks

Date	Time UTC Start	Finish	Frequency (MHz)	Mode	Power	Station	Report Sent	Received	Signal Sent	Received

Remarks

Date	Time UTC Start	Finish	Frequency (MHz)	Mode	Power	Station	Report Sent	Received	Signal Sent	Received

Remarks

Date	Time UTC Start	Finish	Frequency (MHz)	Mode	Power	Station	Report Sent	Received	Signal Sent	Received

Remarks

AMATEUR RADIO STATION LOG

Date	Time UTC		Frequency (MHz)	Mode	Power	Station	Report		Signal	
	Start	Finish					Sent	Received	Sent	Received

Remarks

Date	Time UTC		Frequency (MHz)	Mode	Power	Station	Report		Signal	
	Start	Finish					Sent	Received	Sent	Received

Remarks

Date	Time UTC		Frequency (MHz)	Mode	Power	Station	Report		Signal	
	Start	Finish					Sent	Received	Sent	Received

Remarks

Date	Time UTC		Frequency (MHz)	Mode	Power	Station	Report		Signal	
	Start	Finish					Sent	Received	Sent	Received

Remarks

Date	Time UTC		Frequency (MHz)	Mode	Power	Station	Report		Signal	
	Start	Finish					Sent	Received	Sent	Received

Remarks

Date	Time UTC		Frequency (MHz)	Mode	Power	Station	Report		Signal	
	Start	Finish					Sent	Received	Sent	Received

Remarks

Date	Time UTC		Frequency (MHz)	Mode	Power	Station	Report		Signal	
	Start	Finish					Sent	Received	Sent	Received

Remarks

AMATEUR RADIO STATION LOG

Date	Time UTC Start	Finish	Frequency (MHz)	Mode	Power	Station	Report Sent	Received	Signal Sent	Received

Remarks

Date	Time UTC Start	Finish	Frequency (MHz)	Mode	Power	Station	Report Sent	Received	Signal Sent	Received

Remarks

Date	Time UTC Start	Finish	Frequency (MHz)	Mode	Power	Station	Report Sent	Received	Signal Sent	Received

Remarks

Date	Time UTC Start	Finish	Frequency (MHz)	Mode	Power	Station	Report Sent	Received	Signal Sent	Received

Remarks

Date	Time UTC Start	Finish	Frequency (MHz)	Mode	Power	Station	Report Sent	Received	Signal Sent	Received

Remarks

Date	Time UTC Start	Finish	Frequency (MHz)	Mode	Power	Station	Report Sent	Received	Signal Sent	Received

Remarks

Date	Time UTC Start	Finish	Frequency (MHz)	Mode	Power	Station	Report Sent	Received	Signal Sent	Received

Remarks

AMATEUR RADIO STATION LOG

Date	Time UTC Start	Time UTC Finish	Frequency (MHz)	Mode	Power	Station	Report Sent	Report Received	Signal Sent	Signal Received

Remarks

Date	Time UTC Start	Time UTC Finish	Frequency (MHz)	Mode	Power	Station	Report Sent	Report Received	Signal Sent	Signal Received

Remarks

Date	Time UTC Start	Time UTC Finish	Frequency (MHz)	Mode	Power	Station	Report Sent	Report Received	Signal Sent	Signal Received

Remarks

Date	Time UTC Start	Time UTC Finish	Frequency (MHz)	Mode	Power	Station	Report Sent	Report Received	Signal Sent	Signal Received

Remarks

Date	Time UTC Start	Time UTC Finish	Frequency (MHz)	Mode	Power	Station	Report Sent	Report Received	Signal Sent	Signal Received

Remarks

Date	Time UTC Start	Time UTC Finish	Frequency (MHz)	Mode	Power	Station	Report Sent	Report Received	Signal Sent	Signal Received

Remarks

Date	Time UTC Start	Time UTC Finish	Frequency (MHz)	Mode	Power	Station	Report Sent	Report Received	Signal Sent	Signal Received

Remarks

AMATEUR RADIO STATION LOG

Date	Time UTC Start	Time UTC Finish	Frequency (MHz)	Mode	Power	Station	Report Sent	Report Received	Signal Sent	Signal Received

Remarks

Date	Time UTC Start	Time UTC Finish	Frequency (MHz)	Mode	Power	Station	Report Sent	Report Received	Signal Sent	Signal Received

Remarks

Date	Time UTC Start	Time UTC Finish	Frequency (MHz)	Mode	Power	Station	Report Sent	Report Received	Signal Sent	Signal Received

Remarks

Date	Time UTC Start	Time UTC Finish	Frequency (MHz)	Mode	Power	Station	Report Sent	Report Received	Signal Sent	Signal Received

Remarks

Date	Time UTC Start	Time UTC Finish	Frequency (MHz)	Mode	Power	Station	Report Sent	Report Received	Signal Sent	Signal Received

Remarks

Date	Time UTC Start	Time UTC Finish	Frequency (MHz)	Mode	Power	Station	Report Sent	Report Received	Signal Sent	Signal Received

Remarks

Date	Time UTC Start	Time UTC Finish	Frequency (MHz)	Mode	Power	Station	Report Sent	Report Received	Signal Sent	Signal Received

Remarks

AMATEUR RADIO STATION LOG

Date	Time UTC Start	Finish	Frequency (MHz)	Mode	Power	Station	Report Sent	Received	Signal Sent	Received

Remarks

Date	Time UTC Start	Finish	Frequency (MHz)	Mode	Power	Station	Report Sent	Received	Signal Sent	Received

Remarks

Date	Time UTC Start	Finish	Frequency (MHz)	Mode	Power	Station	Report Sent	Received	Signal Sent	Received

Remarks

Date	Time UTC Start	Finish	Frequency (MHz)	Mode	Power	Station	Report Sent	Received	Signal Sent	Received

Remarks

Date	Time UTC Start	Finish	Frequency (MHz)	Mode	Power	Station	Report Sent	Received	Signal Sent	Received

Remarks

Date	Time UTC Start	Finish	Frequency (MHz)	Mode	Power	Station	Report Sent	Received	Signal Sent	Received

Remarks

Date	Time UTC Start	Finish	Frequency (MHz)	Mode	Power	Station	Report Sent	Received	Signal Sent	Received

Remarks

AMATEUR RADIO STATION LOG

Date	Time UTC Start	Time UTC Finish	Frequency (MHz)	Mode	Power	Station	Report Sent	Report Received	Signal Sent	Signal Received

Remarks

Date	Time UTC Start	Time UTC Finish	Frequency (MHz)	Mode	Power	Station	Report Sent	Report Received	Signal Sent	Signal Received

Remarks

Date	Time UTC Start	Time UTC Finish	Frequency (MHz)	Mode	Power	Station	Report Sent	Report Received	Signal Sent	Signal Received

Remarks

Date	Time UTC Start	Time UTC Finish	Frequency (MHz)	Mode	Power	Station	Report Sent	Report Received	Signal Sent	Signal Received

Remarks

Date	Time UTC Start	Time UTC Finish	Frequency (MHz)	Mode	Power	Station	Report Sent	Report Received	Signal Sent	Signal Received

Remarks

Date	Time UTC Start	Time UTC Finish	Frequency (MHz)	Mode	Power	Station	Report Sent	Report Received	Signal Sent	Signal Received

Remarks

Date	Time UTC Start	Time UTC Finish	Frequency (MHz)	Mode	Power	Station	Report Sent	Report Received	Signal Sent	Signal Received

Remarks

AMATEUR RADIO STATION LOG

Date	Time UTC Start	Finish	Frequency (MHz)	Mode	Power	Station	Report Sent	Received	Signal Sent	Received

Remarks

Date	Time UTC Start	Finish	Frequency (MHz)	Mode	Power	Station	Report Sent	Received	Signal Sent	Received

Remarks

Date	Time UTC Start	Finish	Frequency (MHz)	Mode	Power	Station	Report Sent	Received	Signal Sent	Received

Remarks

Date	Time UTC Start	Finish	Frequency (MHz)	Mode	Power	Station	Report Sent	Received	Signal Sent	Received

Remarks

Date	Time UTC Start	Finish	Frequency (MHz)	Mode	Power	Station	Report Sent	Received	Signal Sent	Received

Remarks

Date	Time UTC Start	Finish	Frequency (MHz)	Mode	Power	Station	Report Sent	Received	Signal Sent	Received

Remarks

Date	Time UTC Start	Finish	Frequency (MHz)	Mode	Power	Station	Report Sent	Received	Signal Sent	Received

Remarks

AMATEUR RADIO STATION LOG

Date	Time UTC Start	Finish	Frequency (MHz)	Mode	Power	Station	Report Sent	Received	Signal Sent	Received

Remarks

Date	Time UTC Start	Finish	Frequency (MHz)	Mode	Power	Station	Report Sent	Received	Signal Sent	Received

Remarks

Date	Time UTC Start	Finish	Frequency (MHz)	Mode	Power	Station	Report Sent	Received	Signal Sent	Received

Remarks

Date	Time UTC Start	Finish	Frequency (MHz)	Mode	Power	Station	Report Sent	Received	Signal Sent	Received

Remarks

Date	Time UTC Start	Finish	Frequency (MHz)	Mode	Power	Station	Report Sent	Received	Signal Sent	Received

Remarks

Date	Time UTC Start	Finish	Frequency (MHz)	Mode	Power	Station	Report Sent	Received	Signal Sent	Received

Remarks

Date	Time UTC Start	Finish	Frequency (MHz)	Mode	Power	Station	Report Sent	Received	Signal Sent	Received

Remarks

AMATEUR RADIO STATION LOG

Date	Time UTC		Frequency (MHz)	Mode	Power	Station	Report		Signal	
	Start	Finish					Sent	Received	Sent	Received

Remarks

Date	Time UTC		Frequency (MHz)	Mode	Power	Station	Report		Signal	
	Start	Finish					Sent	Received	Sent	Received

Remarks

Date	Time UTC		Frequency (MHz)	Mode	Power	Station	Report		Signal	
	Start	Finish					Sent	Received	Sent	Received

Remarks

Date	Time UTC		Frequency (MHz)	Mode	Power	Station	Report		Signal	
	Start	Finish					Sent	Received	Sent	Received

Remarks

Date	Time UTC		Frequency (MHz)	Mode	Power	Station	Report		Signal	
	Start	Finish					Sent	Received	Sent	Received

Remarks

Date	Time UTC		Frequency (MHz)	Mode	Power	Station	Report		Signal	
	Start	Finish					Sent	Received	Sent	Received

Remarks

Date	Time UTC		Frequency (MHz)	Mode	Power	Station	Report		Signal	
	Start	Finish					Sent	Received	Sent	Received

Remarks

AMATEUR RADIO STATION LOG

Date	Time UTC Start	Finish	Frequency (MHz)	Mode	Power	Station	Report Sent	Received	Signal Sent	Received

Remarks

Date	Time UTC Start	Finish	Frequency (MHz)	Mode	Power	Station	Report Sent	Received	Signal Sent	Received

Remarks

Date	Time UTC Start	Finish	Frequency (MHz)	Mode	Power	Station	Report Sent	Received	Signal Sent	Received

Remarks

Date	Time UTC Start	Finish	Frequency (MHz)	Mode	Power	Station	Report Sent	Received	Signal Sent	Received

Remarks

Date	Time UTC Start	Finish	Frequency (MHz)	Mode	Power	Station	Report Sent	Received	Signal Sent	Received

Remarks

Date	Time UTC Start	Finish	Frequency (MHz)	Mode	Power	Station	Report Sent	Received	Signal Sent	Received

Remarks

Date	Time UTC Start	Finish	Frequency (MHz)	Mode	Power	Station	Report Sent	Received	Signal Sent	Received

Remarks

AMATEUR RADIO STATION LOG

Date	Time UTC		Frequency (MHz)	Mode	Power	Station	Report		Signal	
	Start	Finish					Sent	Received	Sent	Received

Remarks

Date	Time UTC		Frequency (MHz)	Mode	Power	Station	Report		Signal	
	Start	Finish					Sent	Received	Sent	Received

Remarks

Date	Time UTC		Frequency (MHz)	Mode	Power	Station	Report		Signal	
	Start	Finish					Sent	Received	Sent	Received

Remarks

Date	Time UTC		Frequency (MHz)	Mode	Power	Station	Report		Signal	
	Start	Finish					Sent	Received	Sent	Received

Remarks

Date	Time UTC		Frequency (MHz)	Mode	Power	Station	Report		Signal	
	Start	Finish					Sent	Received	Sent	Received

Remarks

Date	Time UTC		Frequency (MHz)	Mode	Power	Station	Report		Signal	
	Start	Finish					Sent	Received	Sent	Received

Remarks

Date	Time UTC		Frequency (MHz)	Mode	Power	Station	Report		Signal	
	Start	Finish					Sent	Received	Sent	Received

Remarks

AMATEUR RADIO STATION LOG

Date	Time UTC Start	Finish	Frequency (MHz)	Mode	Power	Station	Report Sent	Received	Signal Sent	Received

Remarks

Date	Time UTC Start	Finish	Frequency (MHz)	Mode	Power	Station	Report Sent	Received	Signal Sent	Received

Remarks

Date	Time UTC Start	Finish	Frequency (MHz)	Mode	Power	Station	Report Sent	Received	Signal Sent	Received

Remarks

Date	Time UTC Start	Finish	Frequency (MHz)	Mode	Power	Station	Report Sent	Received	Signal Sent	Received

Remarks

Date	Time UTC Start	Finish	Frequency (MHz)	Mode	Power	Station	Report Sent	Received	Signal Sent	Received

Remarks

Date	Time UTC Start	Finish	Frequency (MHz)	Mode	Power	Station	Report Sent	Received	Signal Sent	Received

Remarks

Date	Time UTC Start	Finish	Frequency (MHz)	Mode	Power	Station	Report Sent	Received	Signal Sent	Received

Remarks

AMATEUR RADIO STATION LOG

Date	Time UTC Start	Finish	Frequency (MHz)	Mode	Power	Station	Report Sent	Received	Signal Sent	Received

Remarks

Date	Time UTC Start	Finish	Frequency (MHz)	Mode	Power	Station	Report Sent	Received	Signal Sent	Received

Remarks

Date	Time UTC Start	Finish	Frequency (MHz)	Mode	Power	Station	Report Sent	Received	Signal Sent	Received

Remarks

Date	Time UTC Start	Finish	Frequency (MHz)	Mode	Power	Station	Report Sent	Received	Signal Sent	Received

Remarks

Date	Time UTC Start	Finish	Frequency (MHz)	Mode	Power	Station	Report Sent	Received	Signal Sent	Received

Remarks

Date	Time UTC Start	Finish	Frequency (MHz)	Mode	Power	Station	Report Sent	Received	Signal Sent	Received

Remarks

Date	Time UTC Start	Finish	Frequency (MHz)	Mode	Power	Station	Report Sent	Received	Signal Sent	Received

Remarks

AMATEUR RADIO STATION LOG

Date	Time UTC Start	Finish	Frequency (MHz)	Mode	Power	Station	Report Sent	Received	Signal Sent	Received
Remarks										

Date	Time UTC Start	Finish	Frequency (MHz)	Mode	Power	Station	Report Sent	Received	Signal Sent	Received
Remarks										

Date	Time UTC Start	Finish	Frequency (MHz)	Mode	Power	Station	Report Sent	Received	Signal Sent	Received
Remarks										

Date	Time UTC Start	Finish	Frequency (MHz)	Mode	Power	Station	Report Sent	Received	Signal Sent	Received
Remarks										

Date	Time UTC Start	Finish	Frequency (MHz)	Mode	Power	Station	Report Sent	Received	Signal Sent	Received
Remarks										

Date	Time UTC Start	Finish	Frequency (MHz)	Mode	Power	Station	Report Sent	Received	Signal Sent	Received
Remarks										

Date	Time UTC Start	Finish	Frequency (MHz)	Mode	Power	Station	Report Sent	Received	Signal Sent	Received
Remarks										

AMATEUR RADIO STATION LOG

Date	Time UTC Start	Finish	Frequency (MHz)	Mode	Power	Station	Report Sent	Received	Signal Sent	Received

Remarks

Date	Time UTC Start	Finish	Frequency (MHz)	Mode	Power	Station	Report Sent	Received	Signal Sent	Received

Remarks

Date	Time UTC Start	Finish	Frequency (MHz)	Mode	Power	Station	Report Sent	Received	Signal Sent	Received

Remarks

Date	Time UTC Start	Finish	Frequency (MHz)	Mode	Power	Station	Report Sent	Received	Signal Sent	Received

Remarks

Date	Time UTC Start	Finish	Frequency (MHz)	Mode	Power	Station	Report Sent	Received	Signal Sent	Received

Remarks

Date	Time UTC Start	Finish	Frequency (MHz)	Mode	Power	Station	Report Sent	Received	Signal Sent	Received

Remarks

Date	Time UTC Start	Finish	Frequency (MHz)	Mode	Power	Station	Report Sent	Received	Signal Sent	Received

Remarks

Amateur Radio Station Log

Date	Time UTC Start	Time UTC Finish	Frequency (MHz)	Mode	Power	Station	Report Sent	Report Received	Signal Sent	Signal Received
Remarks										

Date	Time UTC Start	Time UTC Finish	Frequency (MHz)	Mode	Power	Station	Report Sent	Report Received	Signal Sent	Signal Received
Remarks										

Date	Time UTC Start	Time UTC Finish	Frequency (MHz)	Mode	Power	Station	Report Sent	Report Received	Signal Sent	Signal Received
Remarks										

Date	Time UTC Start	Time UTC Finish	Frequency (MHz)	Mode	Power	Station	Report Sent	Report Received	Signal Sent	Signal Received
Remarks										

Date	Time UTC Start	Time UTC Finish	Frequency (MHz)	Mode	Power	Station	Report Sent	Report Received	Signal Sent	Signal Received
Remarks										

Date	Time UTC Start	Time UTC Finish	Frequency (MHz)	Mode	Power	Station	Report Sent	Report Received	Signal Sent	Signal Received
Remarks										

Date	Time UTC Start	Time UTC Finish	Frequency (MHz)	Mode	Power	Station	Report Sent	Report Received	Signal Sent	Signal Received
Remarks										

AMATEUR RADIO STATION LOG

Date	Time UTC		Frequency (MHz)	Mode	Power	Station	Report		Signal	
	Start	Finish					Sent	Received	Sent	Received

Remarks

Date	Time UTC		Frequency (MHz)	Mode	Power	Station	Report		Signal	
	Start	Finish					Sent	Received	Sent	Received

Remarks

Date	Time UTC		Frequency (MHz)	Mode	Power	Station	Report		Signal	
	Start	Finish					Sent	Received	Sent	Received

Remarks

Date	Time UTC		Frequency (MHz)	Mode	Power	Station	Report		Signal	
	Start	Finish					Sent	Received	Sent	Received

Remarks

Date	Time UTC		Frequency (MHz)	Mode	Power	Station	Report		Signal	
	Start	Finish					Sent	Received	Sent	Received

Remarks

Date	Time UTC		Frequency (MHz)	Mode	Power	Station	Report		Signal	
	Start	Finish					Sent	Received	Sent	Received

Remarks

Date	Time UTC		Frequency (MHz)	Mode	Power	Station	Report		Signal	
	Start	Finish					Sent	Received	Sent	Received

Remarks

AMATEUR RADIO STATION LOG

Date	Time UTC Start	Time UTC Finish	Frequency (MHz)	Mode	Power	Station	Report Sent	Report Received	Signal Sent	Signal Received

Remarks

Date	Time UTC Start	Time UTC Finish	Frequency (MHz)	Mode	Power	Station	Report Sent	Report Received	Signal Sent	Signal Received

Remarks

Date	Time UTC Start	Time UTC Finish	Frequency (MHz)	Mode	Power	Station	Report Sent	Report Received	Signal Sent	Signal Received

Remarks

Date	Time UTC Start	Time UTC Finish	Frequency (MHz)	Mode	Power	Station	Report Sent	Report Received	Signal Sent	Signal Received

Remarks

Date	Time UTC Start	Time UTC Finish	Frequency (MHz)	Mode	Power	Station	Report Sent	Report Received	Signal Sent	Signal Received

Remarks

Date	Time UTC Start	Time UTC Finish	Frequency (MHz)	Mode	Power	Station	Report Sent	Report Received	Signal Sent	Signal Received

Remarks

Date	Time UTC Start	Time UTC Finish	Frequency (MHz)	Mode	Power	Station	Report Sent	Report Received	Signal Sent	Signal Received

Remarks

AMATEUR RADIO STATION LOG

Date	Time UTC		Frequency (MHz)	Mode	Power	Station	Report		Signal	
	Start	Finish					Sent	Received	Sent	Received

Remarks

Date	Time UTC		Frequency (MHz)	Mode	Power	Station	Report		Signal	
	Start	Finish					Sent	Received	Sent	Received

Remarks

Date	Time UTC		Frequency (MHz)	Mode	Power	Station	Report		Signal	
	Start	Finish					Sent	Received	Sent	Received

Remarks

Date	Time UTC		Frequency (MHz)	Mode	Power	Station	Report		Signal	
	Start	Finish					Sent	Received	Sent	Received

Remarks

Date	Time UTC		Frequency (MHz)	Mode	Power	Station	Report		Signal	
	Start	Finish					Sent	Received	Sent	Received

Remarks

Date	Time UTC		Frequency (MHz)	Mode	Power	Station	Report		Signal	
	Start	Finish					Sent	Received	Sent	Received

Remarks

Date	Time UTC		Frequency (MHz)	Mode	Power	Station	Report		Signal	
	Start	Finish					Sent	Received	Sent	Received

Remarks

AMATEUR RADIO STATION LOG

Date	Time UTC Start	Finish	Frequency (MHz)	Mode	Power	Station	Report Sent	Received	Signal Sent	Received

Remarks

Date	Time UTC Start	Finish	Frequency (MHz)	Mode	Power	Station	Report Sent	Received	Signal Sent	Received

Remarks

Date	Time UTC Start	Finish	Frequency (MHz)	Mode	Power	Station	Report Sent	Received	Signal Sent	Received

Remarks

Date	Time UTC Start	Finish	Frequency (MHz)	Mode	Power	Station	Report Sent	Received	Signal Sent	Received

Remarks

Date	Time UTC Start	Finish	Frequency (MHz)	Mode	Power	Station	Report Sent	Received	Signal Sent	Received

Remarks

Date	Time UTC Start	Finish	Frequency (MHz)	Mode	Power	Station	Report Sent	Received	Signal Sent	Received

Remarks

Date	Time UTC Start	Finish	Frequency (MHz)	Mode	Power	Station	Report Sent	Received	Signal Sent	Received

Remarks

AMATEUR RADIO STATION LOG

Date	Time UTC Start	Finish	Frequency (MHz)	Mode	Power	Station	Report Sent	Received	Signal Sent	Received

Remarks

Date	Time UTC Start	Finish	Frequency (MHz)	Mode	Power	Station	Report Sent	Received	Signal Sent	Received

Remarks

Date	Time UTC Start	Finish	Frequency (MHz)	Mode	Power	Station	Report Sent	Received	Signal Sent	Received

Remarks

Date	Time UTC Start	Finish	Frequency (MHz)	Mode	Power	Station	Report Sent	Received	Signal Sent	Received

Remarks

Date	Time UTC Start	Finish	Frequency (MHz)	Mode	Power	Station	Report Sent	Received	Signal Sent	Received

Remarks

Date	Time UTC Start	Finish	Frequency (MHz)	Mode	Power	Station	Report Sent	Received	Signal Sent	Received

Remarks

Date	Time UTC Start	Finish	Frequency (MHz)	Mode	Power	Station	Report Sent	Received	Signal Sent	Received

Remarks

AMATEUR RADIO STATION LOG

Date	Time UTC Start	Finish	Frequency (MHz)	Mode	Power	Station	Report Sent	Received	Signal Sent	Received
Remarks										

Date	Time UTC Start	Finish	Frequency (MHz)	Mode	Power	Station	Report Sent	Received	Signal Sent	Received
Remarks										

Date	Time UTC Start	Finish	Frequency (MHz)	Mode	Power	Station	Report Sent	Received	Signal Sent	Received
Remarks										

Date	Time UTC Start	Finish	Frequency (MHz)	Mode	Power	Station	Report Sent	Received	Signal Sent	Received
Remarks										

Date	Time UTC Start	Finish	Frequency (MHz)	Mode	Power	Station	Report Sent	Received	Signal Sent	Received
Remarks										

Date	Time UTC Start	Finish	Frequency (MHz)	Mode	Power	Station	Report Sent	Received	Signal Sent	Received
Remarks										

Date	Time UTC Start	Finish	Frequency (MHz)	Mode	Power	Station	Report Sent	Received	Signal Sent	Received
Remarks										

AMATEUR RADIO STATION LOG

Date	Time UTC Start	Finish	Frequency (MHz)	Mode	Power	Station	Report Sent	Received	Signal Sent	Received
Remarks										

Date	Time UTC Start	Finish	Frequency (MHz)	Mode	Power	Station	Report Sent	Received	Signal Sent	Received
Remarks										

Date	Time UTC Start	Finish	Frequency (MHz)	Mode	Power	Station	Report Sent	Received	Signal Sent	Received
Remarks										

Date	Time UTC Start	Finish	Frequency (MHz)	Mode	Power	Station	Report Sent	Received	Signal Sent	Received
Remarks										

Date	Time UTC Start	Finish	Frequency (MHz)	Mode	Power	Station	Report Sent	Received	Signal Sent	Received
Remarks										

Date	Time UTC Start	Finish	Frequency (MHz)	Mode	Power	Station	Report Sent	Received	Signal Sent	Received
Remarks										

Date	Time UTC Start	Finish	Frequency (MHz)	Mode	Power	Station	Report Sent	Received	Signal Sent	Received
Remarks										

AMATEUR RADIO STATION LOG

Date	Time UTC		Frequency (MHz)	Mode	Power	Station	Report		Signal	
	Start	Finish					Sent	Received	Sent	Received

Remarks

Date	Time UTC		Frequency (MHz)	Mode	Power	Station	Report		Signal	
	Start	Finish					Sent	Received	Sent	Received

Remarks

Date	Time UTC		Frequency (MHz)	Mode	Power	Station	Report		Signal	
	Start	Finish					Sent	Received	Sent	Received

Remarks

Date	Time UTC		Frequency (MHz)	Mode	Power	Station	Report		Signal	
	Start	Finish					Sent	Received	Sent	Received

Remarks

Date	Time UTC		Frequency (MHz)	Mode	Power	Station	Report		Signal	
	Start	Finish					Sent	Received	Sent	Received

Remarks

Date	Time UTC		Frequency (MHz)	Mode	Power	Station	Report		Signal	
	Start	Finish					Sent	Received	Sent	Received

Remarks

Date	Time UTC		Frequency (MHz)	Mode	Power	Station	Report		Signal	
	Start	Finish					Sent	Received	Sent	Received

Remarks

AMATEUR RADIO STATION LOG

Date	Time UTC		Frequency (MHz)	Mode	Power	Station	Report		Signal	
	Start	Finish					Sent	Received	Sent	Received

Remarks

Date	Time UTC		Frequency (MHz)	Mode	Power	Station	Report		Signal	
	Start	Finish					Sent	Received	Sent	Received

Remarks

Date	Time UTC		Frequency (MHz)	Mode	Power	Station	Report		Signal	
	Start	Finish					Sent	Received	Sent	Received

Remarks

Date	Time UTC		Frequency (MHz)	Mode	Power	Station	Report		Signal	
	Start	Finish					Sent	Received	Sent	Received

Remarks

Date	Time UTC		Frequency (MHz)	Mode	Power	Station	Report		Signal	
	Start	Finish					Sent	Received	Sent	Received

Remarks

Date	Time UTC		Frequency (MHz)	Mode	Power	Station	Report		Signal	
	Start	Finish					Sent	Received	Sent	Received

Remarks

Date	Time UTC		Frequency (MHz)	Mode	Power	Station	Report		Signal	
	Start	Finish					Sent	Received	Sent	Received

Remarks

AMATEUR RADIO STATION LOG

Date	Time UTC Start	Time UTC Finish	Frequency (MHz)	Mode	Power	Station	Report Sent	Report Received	Signal Sent	Signal Received

Remarks

Date	Time UTC Start	Time UTC Finish	Frequency (MHz)	Mode	Power	Station	Report Sent	Report Received	Signal Sent	Signal Received

Remarks

Date	Time UTC Start	Time UTC Finish	Frequency (MHz)	Mode	Power	Station	Report Sent	Report Received	Signal Sent	Signal Received

Remarks

Date	Time UTC Start	Time UTC Finish	Frequency (MHz)	Mode	Power	Station	Report Sent	Report Received	Signal Sent	Signal Received

Remarks

Date	Time UTC Start	Time UTC Finish	Frequency (MHz)	Mode	Power	Station	Report Sent	Report Received	Signal Sent	Signal Received

Remarks

Date	Time UTC Start	Time UTC Finish	Frequency (MHz)	Mode	Power	Station	Report Sent	Report Received	Signal Sent	Signal Received

Remarks

Date	Time UTC Start	Time UTC Finish	Frequency (MHz)	Mode	Power	Station	Report Sent	Report Received	Signal Sent	Signal Received

Remarks

AMATEUR RADIO STATION LOG

Date	Time UTC Start	Finish	Frequency (MHz)	Mode	Power	Station	Report Sent	Received	Signal Sent	Received

Remarks

Date	Time UTC Start	Finish	Frequency (MHz)	Mode	Power	Station	Report Sent	Received	Signal Sent	Received

Remarks

Date	Time UTC Start	Finish	Frequency (MHz)	Mode	Power	Station	Report Sent	Received	Signal Sent	Received

Remarks

Date	Time UTC Start	Finish	Frequency (MHz)	Mode	Power	Station	Report Sent	Received	Signal Sent	Received

Remarks

Date	Time UTC Start	Finish	Frequency (MHz)	Mode	Power	Station	Report Sent	Received	Signal Sent	Received

Remarks

Date	Time UTC Start	Finish	Frequency (MHz)	Mode	Power	Station	Report Sent	Received	Signal Sent	Received

Remarks

Date	Time UTC Start	Finish	Frequency (MHz)	Mode	Power	Station	Report Sent	Received	Signal Sent	Received

Remarks

AMATEUR RADIO STATION LOG

Date	Time UTC Start	Time UTC Finish	Frequency (MHz)	Mode	Power	Station	Report Sent	Report Received	Signal Sent	Signal Received

Remarks

Date	Time UTC Start	Time UTC Finish	Frequency (MHz)	Mode	Power	Station	Report Sent	Report Received	Signal Sent	Signal Received

Remarks

Date	Time UTC Start	Time UTC Finish	Frequency (MHz)	Mode	Power	Station	Report Sent	Report Received	Signal Sent	Signal Received

Remarks

Date	Time UTC Start	Time UTC Finish	Frequency (MHz)	Mode	Power	Station	Report Sent	Report Received	Signal Sent	Signal Received

Remarks

Date	Time UTC Start	Time UTC Finish	Frequency (MHz)	Mode	Power	Station	Report Sent	Report Received	Signal Sent	Signal Received

Remarks

Date	Time UTC Start	Time UTC Finish	Frequency (MHz)	Mode	Power	Station	Report Sent	Report Received	Signal Sent	Signal Received

Remarks

Date	Time UTC Start	Time UTC Finish	Frequency (MHz)	Mode	Power	Station	Report Sent	Report Received	Signal Sent	Signal Received

Remarks

AMATEUR RADIO STATION LOG

Date	Time UTC Start	Finish	Frequency (MHz)	Mode	Power	Station	Report Sent	Received	Signal Sent	Received

Remarks

Date	Time UTC Start	Finish	Frequency (MHz)	Mode	Power	Station	Report Sent	Received	Signal Sent	Received

Remarks

Date	Time UTC Start	Finish	Frequency (MHz)	Mode	Power	Station	Report Sent	Received	Signal Sent	Received

Remarks

Date	Time UTC Start	Finish	Frequency (MHz)	Mode	Power	Station	Report Sent	Received	Signal Sent	Received

Remarks

Date	Time UTC Start	Finish	Frequency (MHz)	Mode	Power	Station	Report Sent	Received	Signal Sent	Received

Remarks

Date	Time UTC Start	Finish	Frequency (MHz)	Mode	Power	Station	Report Sent	Received	Signal Sent	Received

Remarks

Date	Time UTC Start	Finish	Frequency (MHz)	Mode	Power	Station	Report Sent	Received	Signal Sent	Received

Remarks

AMATEUR RADIO STATION LOG

Date	Time UTC Start	Time UTC Finish	Frequency (MHz)	Mode	Power	Station	Report Sent	Report Received	Signal Sent	Signal Received

Remarks

Date	Time UTC Start	Time UTC Finish	Frequency (MHz)	Mode	Power	Station	Report Sent	Report Received	Signal Sent	Signal Received

Remarks

Date	Time UTC Start	Time UTC Finish	Frequency (MHz)	Mode	Power	Station	Report Sent	Report Received	Signal Sent	Signal Received

Remarks

Date	Time UTC Start	Time UTC Finish	Frequency (MHz)	Mode	Power	Station	Report Sent	Report Received	Signal Sent	Signal Received

Remarks

Date	Time UTC Start	Time UTC Finish	Frequency (MHz)	Mode	Power	Station	Report Sent	Report Received	Signal Sent	Signal Received

Remarks

Date	Time UTC Start	Time UTC Finish	Frequency (MHz)	Mode	Power	Station	Report Sent	Report Received	Signal Sent	Signal Received

Remarks

Date	Time UTC Start	Time UTC Finish	Frequency (MHz)	Mode	Power	Station	Report Sent	Report Received	Signal Sent	Signal Received

Remarks

AMATEUR RADIO STATION LOG

Date	Time UTC		Frequency (MHz)	Mode	Power	Station	Report		Signal	
	Start	Finish					Sent	Received	Sent	Received

Remarks

Date	Time UTC		Frequency (MHz)	Mode	Power	Station	Report		Signal	
	Start	Finish					Sent	Received	Sent	Received

Remarks

Date	Time UTC		Frequency (MHz)	Mode	Power	Station	Report		Signal	
	Start	Finish					Sent	Received	Sent	Received

Remarks

Date	Time UTC		Frequency (MHz)	Mode	Power	Station	Report		Signal	
	Start	Finish					Sent	Received	Sent	Received

Remarks

Date	Time UTC		Frequency (MHz)	Mode	Power	Station	Report		Signal	
	Start	Finish					Sent	Received	Sent	Received

Remarks

Date	Time UTC		Frequency (MHz)	Mode	Power	Station	Report		Signal	
	Start	Finish					Sent	Received	Sent	Received

Remarks

Date	Time UTC		Frequency (MHz)	Mode	Power	Station	Report		Signal	
	Start	Finish					Sent	Received	Sent	Received

Remarks

AMATEUR RADIO STATION LOG

Date	Time UTC Start	Time UTC Finish	Frequency (MHz)	Mode	Power	Station	Report Sent	Report Received	Signal Sent	Signal Received

Remarks

Date	Time UTC Start	Time UTC Finish	Frequency (MHz)	Mode	Power	Station	Report Sent	Report Received	Signal Sent	Signal Received

Remarks

Date	Time UTC Start	Time UTC Finish	Frequency (MHz)	Mode	Power	Station	Report Sent	Report Received	Signal Sent	Signal Received

Remarks

Date	Time UTC Start	Time UTC Finish	Frequency (MHz)	Mode	Power	Station	Report Sent	Report Received	Signal Sent	Signal Received

Remarks

Date	Time UTC Start	Time UTC Finish	Frequency (MHz)	Mode	Power	Station	Report Sent	Report Received	Signal Sent	Signal Received

Remarks

Date	Time UTC Start	Time UTC Finish	Frequency (MHz)	Mode	Power	Station	Report Sent	Report Received	Signal Sent	Signal Received

Remarks

Date	Time UTC Start	Time UTC Finish	Frequency (MHz)	Mode	Power	Station	Report Sent	Report Received	Signal Sent	Signal Received

Remarks

AMATEUR RADIO STATION LOG

Date	Time UTC Start	Finish	Frequency (MHz)	Mode	Power	Station	Report Sent	Received	Signal Sent	Received

Remarks

Date	Time UTC Start	Finish	Frequency (MHz)	Mode	Power	Station	Report Sent	Received	Signal Sent	Received

Remarks

Date	Time UTC Start	Finish	Frequency (MHz)	Mode	Power	Station	Report Sent	Received	Signal Sent	Received

Remarks

Date	Time UTC Start	Finish	Frequency (MHz)	Mode	Power	Station	Report Sent	Received	Signal Sent	Received

Remarks

Date	Time UTC Start	Finish	Frequency (MHz)	Mode	Power	Station	Report Sent	Received	Signal Sent	Received

Remarks

Date	Time UTC Start	Finish	Frequency (MHz)	Mode	Power	Station	Report Sent	Received	Signal Sent	Received

Remarks

Date	Time UTC Start	Finish	Frequency (MHz)	Mode	Power	Station	Report Sent	Received	Signal Sent	Received

Remarks

AMATEUR RADIO STATION LOG

Date	Time UTC Start	Time UTC Finish	Frequency (MHz)	Mode	Power	Station	Report Sent	Report Received	Signal Sent	Signal Received

Remarks

Date	Time UTC Start	Time UTC Finish	Frequency (MHz)	Mode	Power	Station	Report Sent	Report Received	Signal Sent	Signal Received

Remarks

Date	Time UTC Start	Time UTC Finish	Frequency (MHz)	Mode	Power	Station	Report Sent	Report Received	Signal Sent	Signal Received

Remarks

Date	Time UTC Start	Time UTC Finish	Frequency (MHz)	Mode	Power	Station	Report Sent	Report Received	Signal Sent	Signal Received

Remarks

Date	Time UTC Start	Time UTC Finish	Frequency (MHz)	Mode	Power	Station	Report Sent	Report Received	Signal Sent	Signal Received

Remarks

Date	Time UTC Start	Time UTC Finish	Frequency (MHz)	Mode	Power	Station	Report Sent	Report Received	Signal Sent	Signal Received

Remarks

Date	Time UTC Start	Time UTC Finish	Frequency (MHz)	Mode	Power	Station	Report Sent	Report Received	Signal Sent	Signal Received

Remarks

AMATEUR RADIO STATION LOG

Date	Time UTC		Frequency (MHz)	Mode	Power	Station	Report		Signal	
	Start	Finish					Sent	Received	Sent	Received

Remarks

Date	Time UTC		Frequency (MHz)	Mode	Power	Station	Report		Signal	
	Start	Finish					Sent	Received	Sent	Received

Remarks

Date	Time UTC		Frequency (MHz)	Mode	Power	Station	Report		Signal	
	Start	Finish					Sent	Received	Sent	Received

Remarks

Date	Time UTC		Frequency (MHz)	Mode	Power	Station	Report		Signal	
	Start	Finish					Sent	Received	Sent	Received

Remarks

Date	Time UTC		Frequency (MHz)	Mode	Power	Station	Report		Signal	
	Start	Finish					Sent	Received	Sent	Received

Remarks

Date	Time UTC		Frequency (MHz)	Mode	Power	Station	Report		Signal	
	Start	Finish					Sent	Received	Sent	Received

Remarks

Date	Time UTC		Frequency (MHz)	Mode	Power	Station	Report		Signal	
	Start	Finish					Sent	Received	Sent	Received

Remarks

AMATEUR RADIO STATION LOG

Date	Time UTC Start	Finish	Frequency (MHz)	Mode	Power	Station	Report Sent	Received	Signal Sent	Received

Remarks

Date	Time UTC Start	Finish	Frequency (MHz)	Mode	Power	Station	Report Sent	Received	Signal Sent	Received

Remarks

Date	Time UTC Start	Finish	Frequency (MHz)	Mode	Power	Station	Report Sent	Received	Signal Sent	Received

Remarks

Date	Time UTC Start	Finish	Frequency (MHz)	Mode	Power	Station	Report Sent	Received	Signal Sent	Received

Remarks

Date	Time UTC Start	Finish	Frequency (MHz)	Mode	Power	Station	Report Sent	Received	Signal Sent	Received

Remarks

Date	Time UTC Start	Finish	Frequency (MHz)	Mode	Power	Station	Report Sent	Received	Signal Sent	Received

Remarks

Date	Time UTC Start	Finish	Frequency (MHz)	Mode	Power	Station	Report Sent	Received	Signal Sent	Received

Remarks

AMATEUR RADIO STATION LOG

Date	Time UTC Start	Finish	Frequency (MHz)	Mode	Power	Station	Report Sent	Received	Signal Sent	Received

Remarks

Date	Time UTC Start	Finish	Frequency (MHz)	Mode	Power	Station	Report Sent	Received	Signal Sent	Received

Remarks

Date	Time UTC Start	Finish	Frequency (MHz)	Mode	Power	Station	Report Sent	Received	Signal Sent	Received

Remarks

Date	Time UTC Start	Finish	Frequency (MHz)	Mode	Power	Station	Report Sent	Received	Signal Sent	Received

Remarks

Date	Time UTC Start	Finish	Frequency (MHz)	Mode	Power	Station	Report Sent	Received	Signal Sent	Received

Remarks

Date	Time UTC Start	Finish	Frequency (MHz)	Mode	Power	Station	Report Sent	Received	Signal Sent	Received

Remarks

Date	Time UTC Start	Finish	Frequency (MHz)	Mode	Power	Station	Report Sent	Received	Signal Sent	Received

Remarks

AMATEUR RADIO STATION LOG

Date	Time UTC Start	Time UTC Finish	Frequency (MHz)	Mode	Power	Station	Report Sent	Report Received	Signal Sent	Signal Received

Remarks

Date	Time UTC Start	Time UTC Finish	Frequency (MHz)	Mode	Power	Station	Report Sent	Report Received	Signal Sent	Signal Received

Remarks

Date	Time UTC Start	Time UTC Finish	Frequency (MHz)	Mode	Power	Station	Report Sent	Report Received	Signal Sent	Signal Received

Remarks

Date	Time UTC Start	Time UTC Finish	Frequency (MHz)	Mode	Power	Station	Report Sent	Report Received	Signal Sent	Signal Received

Remarks

Date	Time UTC Start	Time UTC Finish	Frequency (MHz)	Mode	Power	Station	Report Sent	Report Received	Signal Sent	Signal Received

Remarks

Date	Time UTC Start	Time UTC Finish	Frequency (MHz)	Mode	Power	Station	Report Sent	Report Received	Signal Sent	Signal Received

Remarks

Date	Time UTC Start	Time UTC Finish	Frequency (MHz)	Mode	Power	Station	Report Sent	Report Received	Signal Sent	Signal Received

Remarks

AMATEUR RADIO STATION LOG

Date	Time UTC		Frequency (MHz)	Mode	Power	Station	Report		Signal	
	Start	Finish					Sent	Received	Sent	Received

Remarks

Date	Time UTC		Frequency (MHz)	Mode	Power	Station	Report		Signal	
	Start	Finish					Sent	Received	Sent	Received

Remarks

Date	Time UTC		Frequency (MHz)	Mode	Power	Station	Report		Signal	
	Start	Finish					Sent	Received	Sent	Received

Remarks

Date	Time UTC		Frequency (MHz)	Mode	Power	Station	Report		Signal	
	Start	Finish					Sent	Received	Sent	Received

Remarks

Date	Time UTC		Frequency (MHz)	Mode	Power	Station	Report		Signal	
	Start	Finish					Sent	Received	Sent	Received

Remarks

Date	Time UTC		Frequency (MHz)	Mode	Power	Station	Report		Signal	
	Start	Finish					Sent	Received	Sent	Received

Remarks

Date	Time UTC		Frequency (MHz)	Mode	Power	Station	Report		Signal	
	Start	Finish					Sent	Received	Sent	Received

Remarks

AMATEUR RADIO STATION LOG

Date	Time UTC Start	Finish	Frequency (MHz)	Mode	Power	Station	Report Sent	Received	Signal Sent	Received

Remarks

Date	Time UTC Start	Finish	Frequency (MHz)	Mode	Power	Station	Report Sent	Received	Signal Sent	Received

Remarks

Date	Time UTC Start	Finish	Frequency (MHz)	Mode	Power	Station	Report Sent	Received	Signal Sent	Received

Remarks

Date	Time UTC Start	Finish	Frequency (MHz)	Mode	Power	Station	Report Sent	Received	Signal Sent	Received

Remarks

Date	Time UTC Start	Finish	Frequency (MHz)	Mode	Power	Station	Report Sent	Received	Signal Sent	Received

Remarks

Date	Time UTC Start	Finish	Frequency (MHz)	Mode	Power	Station	Report Sent	Received	Signal Sent	Received

Remarks

Date	Time UTC Start	Finish	Frequency (MHz)	Mode	Power	Station	Report Sent	Received	Signal Sent	Received

Remarks

AMATEUR RADIO STATION LOG

Date	Time UTC Start	Finish	Frequency (MHz)	Mode	Power	Station	Report Sent	Received	Signal Sent	Received

Remarks

Date	Time UTC Start	Finish	Frequency (MHz)	Mode	Power	Station	Report Sent	Received	Signal Sent	Received

Remarks

Date	Time UTC Start	Finish	Frequency (MHz)	Mode	Power	Station	Report Sent	Received	Signal Sent	Received

Remarks

Date	Time UTC Start	Finish	Frequency (MHz)	Mode	Power	Station	Report Sent	Received	Signal Sent	Received

Remarks

Date	Time UTC Start	Finish	Frequency (MHz)	Mode	Power	Station	Report Sent	Received	Signal Sent	Received

Remarks

Date	Time UTC Start	Finish	Frequency (MHz)	Mode	Power	Station	Report Sent	Received	Signal Sent	Received

Remarks

Date	Time UTC Start	Finish	Frequency (MHz)	Mode	Power	Station	Report Sent	Received	Signal Sent	Received

Remarks

AMATEUR RADIO STATION LOG

Date	Time UTC Start	Finish	Frequency (MHz)	Mode	Power	Station	Report Sent	Received	Signal Sent	Received

Remarks

Date	Time UTC Start	Finish	Frequency (MHz)	Mode	Power	Station	Report Sent	Received	Signal Sent	Received

Remarks

Date	Time UTC Start	Finish	Frequency (MHz)	Mode	Power	Station	Report Sent	Received	Signal Sent	Received

Remarks

Date	Time UTC Start	Finish	Frequency (MHz)	Mode	Power	Station	Report Sent	Received	Signal Sent	Received

Remarks

Date	Time UTC Start	Finish	Frequency (MHz)	Mode	Power	Station	Report Sent	Received	Signal Sent	Received

Remarks

Date	Time UTC Start	Finish	Frequency (MHz)	Mode	Power	Station	Report Sent	Received	Signal Sent	Received

Remarks

Date	Time UTC Start	Finish	Frequency (MHz)	Mode	Power	Station	Report Sent	Received	Signal Sent	Received

Remarks

AMATEUR RADIO STATION LOG

Date	Time UTC Start	Finish	Frequency (MHz)	Mode	Power	Station	Report Sent	Received	Signal Sent	Received

Remarks

Date	Time UTC Start	Finish	Frequency (MHz)	Mode	Power	Station	Report Sent	Received	Signal Sent	Received

Remarks

Date	Time UTC Start	Finish	Frequency (MHz)	Mode	Power	Station	Report Sent	Received	Signal Sent	Received

Remarks

Date	Time UTC Start	Finish	Frequency (MHz)	Mode	Power	Station	Report Sent	Received	Signal Sent	Received

Remarks

Date	Time UTC Start	Finish	Frequency (MHz)	Mode	Power	Station	Report Sent	Received	Signal Sent	Received

Remarks

Date	Time UTC Start	Finish	Frequency (MHz)	Mode	Power	Station	Report Sent	Received	Signal Sent	Received

Remarks

Date	Time UTC Start	Finish	Frequency (MHz)	Mode	Power	Station	Report Sent	Received	Signal Sent	Received

Remarks

AMATEUR RADIO STATION LOG

Date	Time UTC Start	Time UTC Finish	Frequency (MHz)	Mode	Power	Station	Report Sent	Report Received	Signal Sent	Signal Received

Remarks

Date	Time UTC Start	Time UTC Finish	Frequency (MHz)	Mode	Power	Station	Report Sent	Report Received	Signal Sent	Signal Received

Remarks

Date	Time UTC Start	Time UTC Finish	Frequency (MHz)	Mode	Power	Station	Report Sent	Report Received	Signal Sent	Signal Received

Remarks

Date	Time UTC Start	Time UTC Finish	Frequency (MHz)	Mode	Power	Station	Report Sent	Report Received	Signal Sent	Signal Received

Remarks

Date	Time UTC Start	Time UTC Finish	Frequency (MHz)	Mode	Power	Station	Report Sent	Report Received	Signal Sent	Signal Received

Remarks

Date	Time UTC Start	Time UTC Finish	Frequency (MHz)	Mode	Power	Station	Report Sent	Report Received	Signal Sent	Signal Received

Remarks

Date	Time UTC Start	Time UTC Finish	Frequency (MHz)	Mode	Power	Station	Report Sent	Report Received	Signal Sent	Signal Received

Remarks

AMATEUR RADIO STATION LOG

Date	Time UTC Start	Time UTC Finish	Frequency (MHz)	Mode	Power	Station	Report Sent	Report Received	Signal Sent	Signal Received

Remarks

Date	Time UTC Start	Time UTC Finish	Frequency (MHz)	Mode	Power	Station	Report Sent	Report Received	Signal Sent	Signal Received

Remarks

Date	Time UTC Start	Time UTC Finish	Frequency (MHz)	Mode	Power	Station	Report Sent	Report Received	Signal Sent	Signal Received

Remarks

Date	Time UTC Start	Time UTC Finish	Frequency (MHz)	Mode	Power	Station	Report Sent	Report Received	Signal Sent	Signal Received

Remarks

Date	Time UTC Start	Time UTC Finish	Frequency (MHz)	Mode	Power	Station	Report Sent	Report Received	Signal Sent	Signal Received

Remarks

Date	Time UTC Start	Time UTC Finish	Frequency (MHz)	Mode	Power	Station	Report Sent	Report Received	Signal Sent	Signal Received

Remarks

Date	Time UTC Start	Time UTC Finish	Frequency (MHz)	Mode	Power	Station	Report Sent	Report Received	Signal Sent	Signal Received

Remarks

AMATEUR RADIO STATION LOG

Date	Time UTC Start	Finish	Frequency (MHz)	Mode	Power	Station	Report Sent	Received	Signal Sent	Received

Remarks

Date	Time UTC Start	Finish	Frequency (MHz)	Mode	Power	Station	Report Sent	Received	Signal Sent	Received

Remarks

Date	Time UTC Start	Finish	Frequency (MHz)	Mode	Power	Station	Report Sent	Received	Signal Sent	Received

Remarks

Date	Time UTC Start	Finish	Frequency (MHz)	Mode	Power	Station	Report Sent	Received	Signal Sent	Received

Remarks

Date	Time UTC Start	Finish	Frequency (MHz)	Mode	Power	Station	Report Sent	Received	Signal Sent	Received

Remarks

Date	Time UTC Start	Finish	Frequency (MHz)	Mode	Power	Station	Report Sent	Received	Signal Sent	Received

Remarks

Date	Time UTC Start	Finish	Frequency (MHz)	Mode	Power	Station	Report Sent	Received	Signal Sent	Received

Remarks

AMATEUR RADIO STATION LOG

Date	Time UTC Start	Finish	Frequency (MHz)	Mode	Power	Station	Report Sent	Received	Signal Sent	Received

Remarks

Date	Time UTC Start	Finish	Frequency (MHz)	Mode	Power	Station	Report Sent	Received	Signal Sent	Received

Remarks

Date	Time UTC Start	Finish	Frequency (MHz)	Mode	Power	Station	Report Sent	Received	Signal Sent	Received

Remarks

Date	Time UTC Start	Finish	Frequency (MHz)	Mode	Power	Station	Report Sent	Received	Signal Sent	Received

Remarks

Date	Time UTC Start	Finish	Frequency (MHz)	Mode	Power	Station	Report Sent	Received	Signal Sent	Received

Remarks

Date	Time UTC Start	Finish	Frequency (MHz)	Mode	Power	Station	Report Sent	Received	Signal Sent	Received

Remarks

Date	Time UTC Start	Finish	Frequency (MHz)	Mode	Power	Station	Report Sent	Received	Signal Sent	Received

Remarks

AMATEUR RADIO STATION LOG

Date	Time UTC Start	Finish	Frequency (MHz)	Mode	Power	Station	Report Sent	Received	Signal Sent	Received

Remarks

Date	Time UTC Start	Finish	Frequency (MHz)	Mode	Power	Station	Report Sent	Received	Signal Sent	Received

Remarks

Date	Time UTC Start	Finish	Frequency (MHz)	Mode	Power	Station	Report Sent	Received	Signal Sent	Received

Remarks

Date	Time UTC Start	Finish	Frequency (MHz)	Mode	Power	Station	Report Sent	Received	Signal Sent	Received

Remarks

Date	Time UTC Start	Finish	Frequency (MHz)	Mode	Power	Station	Report Sent	Received	Signal Sent	Received

Remarks

Date	Time UTC Start	Finish	Frequency (MHz)	Mode	Power	Station	Report Sent	Received	Signal Sent	Received

Remarks

Date	Time UTC Start	Finish	Frequency (MHz)	Mode	Power	Station	Report Sent	Received	Signal Sent	Received

Remarks

AMATEUR RADIO STATION LOG

Date	Time UTC Start	Time UTC Finish	Frequency (MHz)	Mode	Power	Station	Report Sent	Report Received	Signal Sent	Signal Received

Remarks

Date	Time UTC Start	Time UTC Finish	Frequency (MHz)	Mode	Power	Station	Report Sent	Report Received	Signal Sent	Signal Received

Remarks

Date	Time UTC Start	Time UTC Finish	Frequency (MHz)	Mode	Power	Station	Report Sent	Report Received	Signal Sent	Signal Received

Remarks

Date	Time UTC Start	Time UTC Finish	Frequency (MHz)	Mode	Power	Station	Report Sent	Report Received	Signal Sent	Signal Received

Remarks

Date	Time UTC Start	Time UTC Finish	Frequency (MHz)	Mode	Power	Station	Report Sent	Report Received	Signal Sent	Signal Received

Remarks

Date	Time UTC Start	Time UTC Finish	Frequency (MHz)	Mode	Power	Station	Report Sent	Report Received	Signal Sent	Signal Received

Remarks

Date	Time UTC Start	Time UTC Finish	Frequency (MHz)	Mode	Power	Station	Report Sent	Report Received	Signal Sent	Signal Received

Remarks

AMATEUR RADIO STATION LOG

Date	Time UTC Start	Finish	Frequency (MHz)	Mode	Power	Station	Report Sent	Received	Signal Sent	Received
Remarks										

Date	Time UTC Start	Finish	Frequency (MHz)	Mode	Power	Station	Report Sent	Received	Signal Sent	Received
Remarks										

Date	Time UTC Start	Finish	Frequency (MHz)	Mode	Power	Station	Report Sent	Received	Signal Sent	Received
Remarks										

Date	Time UTC Start	Finish	Frequency (MHz)	Mode	Power	Station	Report Sent	Received	Signal Sent	Received
Remarks										

Date	Time UTC Start	Finish	Frequency (MHz)	Mode	Power	Station	Report Sent	Received	Signal Sent	Received
Remarks										

Date	Time UTC Start	Finish	Frequency (MHz)	Mode	Power	Station	Report Sent	Received	Signal Sent	Received
Remarks										

Date	Time UTC Start	Finish	Frequency (MHz)	Mode	Power	Station	Report Sent	Received	Signal Sent	Received
Remarks										

AMATEUR RADIO STATION LOG

Date	Time UTC		Frequency (MHz)	Mode	Power	Station	Report		Signal	
	Start	Finish					Sent	Received	Sent	Received

Remarks

Date	Time UTC		Frequency (MHz)	Mode	Power	Station	Report		Signal	
	Start	Finish					Sent	Received	Sent	Received

Remarks

Date	Time UTC		Frequency (MHz)	Mode	Power	Station	Report		Signal	
	Start	Finish					Sent	Received	Sent	Received

Remarks

Date	Time UTC		Frequency (MHz)	Mode	Power	Station	Report		Signal	
	Start	Finish					Sent	Received	Sent	Received

Remarks

Date	Time UTC		Frequency (MHz)	Mode	Power	Station	Report		Signal	
	Start	Finish					Sent	Received	Sent	Received

Remarks

Date	Time UTC		Frequency (MHz)	Mode	Power	Station	Report		Signal	
	Start	Finish					Sent	Received	Sent	Received

Remarks

Date	Time UTC		Frequency (MHz)	Mode	Power	Station	Report		Signal	
	Start	Finish					Sent	Received	Sent	Received

Remarks

AMATEUR RADIO STATION LOG

Date	Time UTC Start	Time UTC Finish	Frequency (MHz)	Mode	Power	Station	Report Sent	Report Received	Signal Sent	Signal Received

Remarks

Date	Time UTC Start	Time UTC Finish	Frequency (MHz)	Mode	Power	Station	Report Sent	Report Received	Signal Sent	Signal Received

Remarks

Date	Time UTC Start	Time UTC Finish	Frequency (MHz)	Mode	Power	Station	Report Sent	Report Received	Signal Sent	Signal Received

Remarks

Date	Time UTC Start	Time UTC Finish	Frequency (MHz)	Mode	Power	Station	Report Sent	Report Received	Signal Sent	Signal Received

Remarks

Date	Time UTC Start	Time UTC Finish	Frequency (MHz)	Mode	Power	Station	Report Sent	Report Received	Signal Sent	Signal Received

Remarks

Date	Time UTC Start	Time UTC Finish	Frequency (MHz)	Mode	Power	Station	Report Sent	Report Received	Signal Sent	Signal Received

Remarks

Date	Time UTC Start	Time UTC Finish	Frequency (MHz)	Mode	Power	Station	Report Sent	Report Received	Signal Sent	Signal Received

Remarks

AMATEUR RADIO STATION LOG

Date	Time UTC Start	Finish	Frequency (MHz)	Mode	Power	Station	Report Sent	Received	Signal Sent	Received

Remarks

Date	Time UTC Start	Finish	Frequency (MHz)	Mode	Power	Station	Report Sent	Received	Signal Sent	Received

Remarks

Date	Time UTC Start	Finish	Frequency (MHz)	Mode	Power	Station	Report Sent	Received	Signal Sent	Received

Remarks

Date	Time UTC Start	Finish	Frequency (MHz)	Mode	Power	Station	Report Sent	Received	Signal Sent	Received

Remarks

Date	Time UTC Start	Finish	Frequency (MHz)	Mode	Power	Station	Report Sent	Received	Signal Sent	Received

Remarks

Date	Time UTC Start	Finish	Frequency (MHz)	Mode	Power	Station	Report Sent	Received	Signal Sent	Received

Remarks

Date	Time UTC Start	Finish	Frequency (MHz)	Mode	Power	Station	Report Sent	Received	Signal Sent	Received

Remarks

AMATEUR RADIO STATION LOG

Date	Time UTC		Frequency (MHz)	Mode	Power	Station	Report		Signal	
	Start	Finish					Sent	Received	Sent	Received

Remarks

Date	Time UTC		Frequency (MHz)	Mode	Power	Station	Report		Signal	
	Start	Finish					Sent	Received	Sent	Received

Remarks

Date	Time UTC		Frequency (MHz)	Mode	Power	Station	Report		Signal	
	Start	Finish					Sent	Received	Sent	Received

Remarks

Date	Time UTC		Frequency (MHz)	Mode	Power	Station	Report		Signal	
	Start	Finish					Sent	Received	Sent	Received

Remarks

Date	Time UTC		Frequency (MHz)	Mode	Power	Station	Report		Signal	
	Start	Finish					Sent	Received	Sent	Received

Remarks

Date	Time UTC		Frequency (MHz)	Mode	Power	Station	Report		Signal	
	Start	Finish					Sent	Received	Sent	Received

Remarks

Date	Time UTC		Frequency (MHz)	Mode	Power	Station	Report		Signal	
	Start	Finish					Sent	Received	Sent	Received

Remarks

AMATEUR RADIO STATION LOG

Date	Time UTC Start	Finish	Frequency (MHz)	Mode	Power	Station	Report Sent	Received	Signal Sent	Received

Remarks

Date	Time UTC Start	Finish	Frequency (MHz)	Mode	Power	Station	Report Sent	Received	Signal Sent	Received

Remarks

Date	Time UTC Start	Finish	Frequency (MHz)	Mode	Power	Station	Report Sent	Received	Signal Sent	Received

Remarks

Date	Time UTC Start	Finish	Frequency (MHz)	Mode	Power	Station	Report Sent	Received	Signal Sent	Received

Remarks

Date	Time UTC Start	Finish	Frequency (MHz)	Mode	Power	Station	Report Sent	Received	Signal Sent	Received

Remarks

Date	Time UTC Start	Finish	Frequency (MHz)	Mode	Power	Station	Report Sent	Received	Signal Sent	Received

Remarks

Date	Time UTC Start	Finish	Frequency (MHz)	Mode	Power	Station	Report Sent	Received	Signal Sent	Received

Remarks

AMATEUR RADIO STATION LOG

Date	Time UTC Start	Finish	Frequency (MHz)	Mode	Power	Station	Report Sent	Received	Signal Sent	Received

Remarks

Date	Time UTC Start	Finish	Frequency (MHz)	Mode	Power	Station	Report Sent	Received	Signal Sent	Received

Remarks

Date	Time UTC Start	Finish	Frequency (MHz)	Mode	Power	Station	Report Sent	Received	Signal Sent	Received

Remarks

Date	Time UTC Start	Finish	Frequency (MHz)	Mode	Power	Station	Report Sent	Received	Signal Sent	Received

Remarks

Date	Time UTC Start	Finish	Frequency (MHz)	Mode	Power	Station	Report Sent	Received	Signal Sent	Received

Remarks

Date	Time UTC Start	Finish	Frequency (MHz)	Mode	Power	Station	Report Sent	Received	Signal Sent	Received

Remarks

Date	Time UTC Start	Finish	Frequency (MHz)	Mode	Power	Station	Report Sent	Received	Signal Sent	Received

Remarks

AMATEUR RADIO STATION LOG

Date	Time UTC Start	Finish	Frequency (MHz)	Mode	Power	Station	Report Sent	Received	Signal Sent	Received

Remarks

Date	Time UTC Start	Finish	Frequency (MHz)	Mode	Power	Station	Report Sent	Received	Signal Sent	Received

Remarks

Date	Time UTC Start	Finish	Frequency (MHz)	Mode	Power	Station	Report Sent	Received	Signal Sent	Received

Remarks

Date	Time UTC Start	Finish	Frequency (MHz)	Mode	Power	Station	Report Sent	Received	Signal Sent	Received

Remarks

Date	Time UTC Start	Finish	Frequency (MHz)	Mode	Power	Station	Report Sent	Received	Signal Sent	Received

Remarks

Date	Time UTC Start	Finish	Frequency (MHz)	Mode	Power	Station	Report Sent	Received	Signal Sent	Received

Remarks

Date	Time UTC Start	Finish	Frequency (MHz)	Mode	Power	Station	Report Sent	Received	Signal Sent	Received

Remarks

AMATEUR RADIO STATION LOG

Date	Time UTC Start	Finish	Frequency (MHz)	Mode	Power	Station	Report Sent	Received	Signal Sent	Received

Remarks

Date	Time UTC Start	Finish	Frequency (MHz)	Mode	Power	Station	Report Sent	Received	Signal Sent	Received

Remarks

Date	Time UTC Start	Finish	Frequency (MHz)	Mode	Power	Station	Report Sent	Received	Signal Sent	Received

Remarks

Date	Time UTC Start	Finish	Frequency (MHz)	Mode	Power	Station	Report Sent	Received	Signal Sent	Received

Remarks

Date	Time UTC Start	Finish	Frequency (MHz)	Mode	Power	Station	Report Sent	Received	Signal Sent	Received

Remarks

Date	Time UTC Start	Finish	Frequency (MHz)	Mode	Power	Station	Report Sent	Received	Signal Sent	Received

Remarks

Date	Time UTC Start	Finish	Frequency (MHz)	Mode	Power	Station	Report Sent	Received	Signal Sent	Received

Remarks

AMATEUR RADIO STATION LOG

Date	Time UTC Start	Finish	Frequency (MHz)	Mode	Power	Station	Report Sent	Received	Signal Sent	Received

Remarks

Date	Time UTC Start	Finish	Frequency (MHz)	Mode	Power	Station	Report Sent	Received	Signal Sent	Received

Remarks

Date	Time UTC Start	Finish	Frequency (MHz)	Mode	Power	Station	Report Sent	Received	Signal Sent	Received

Remarks

Date	Time UTC Start	Finish	Frequency (MHz)	Mode	Power	Station	Report Sent	Received	Signal Sent	Received

Remarks

Date	Time UTC Start	Finish	Frequency (MHz)	Mode	Power	Station	Report Sent	Received	Signal Sent	Received

Remarks

Date	Time UTC Start	Finish	Frequency (MHz)	Mode	Power	Station	Report Sent	Received	Signal Sent	Received

Remarks

Date	Time UTC Start	Finish	Frequency (MHz)	Mode	Power	Station	Report Sent	Received	Signal Sent	Received

Remarks

AMATEUR RADIO STATION LOG

Date	Time UTC Start	Time UTC Finish	Frequency (MHz)	Mode	Power	Station	Report Sent	Report Received	Signal Sent	Signal Received

Remarks

Date	Time UTC Start	Time UTC Finish	Frequency (MHz)	Mode	Power	Station	Report Sent	Report Received	Signal Sent	Signal Received

Remarks

Date	Time UTC Start	Time UTC Finish	Frequency (MHz)	Mode	Power	Station	Report Sent	Report Received	Signal Sent	Signal Received

Remarks

Date	Time UTC Start	Time UTC Finish	Frequency (MHz)	Mode	Power	Station	Report Sent	Report Received	Signal Sent	Signal Received

Remarks

Date	Time UTC Start	Time UTC Finish	Frequency (MHz)	Mode	Power	Station	Report Sent	Report Received	Signal Sent	Signal Received

Remarks

Date	Time UTC Start	Time UTC Finish	Frequency (MHz)	Mode	Power	Station	Report Sent	Report Received	Signal Sent	Signal Received

Remarks

Date	Time UTC Start	Time UTC Finish	Frequency (MHz)	Mode	Power	Station	Report Sent	Report Received	Signal Sent	Signal Received

Remarks

AMATEUR RADIO STATION LOG

Date	Time UTC Start	Finish	Frequency (MHz)	Mode	Power	Station	Report Sent	Received	Signal Sent	Received
	Remarks									

Date	Time UTC Start	Finish	Frequency (MHz)	Mode	Power	Station	Report Sent	Received	Signal Sent	Received
	Remarks									

Date	Time UTC Start	Finish	Frequency (MHz)	Mode	Power	Station	Report Sent	Received	Signal Sent	Received
	Remarks									

Date	Time UTC Start	Finish	Frequency (MHz)	Mode	Power	Station	Report Sent	Received	Signal Sent	Received
	Remarks									

Date	Time UTC Start	Finish	Frequency (MHz)	Mode	Power	Station	Report Sent	Received	Signal Sent	Received
	Remarks									

Date	Time UTC Start	Finish	Frequency (MHz)	Mode	Power	Station	Report Sent	Received	Signal Sent	Received
	Remarks									

Date	Time UTC Start	Finish	Frequency (MHz)	Mode	Power	Station	Report Sent	Received	Signal Sent	Received
	Remarks									

AMATEUR RADIO STATION LOG

Date	Time UTC Start	Time UTC Finish	Frequency (MHz)	Mode	Power	Station	Report Sent	Report Received	Signal Sent	Signal Received

Remarks

Date	Time UTC Start	Time UTC Finish	Frequency (MHz)	Mode	Power	Station	Report Sent	Report Received	Signal Sent	Signal Received

Remarks

Date	Time UTC Start	Time UTC Finish	Frequency (MHz)	Mode	Power	Station	Report Sent	Report Received	Signal Sent	Signal Received

Remarks

Date	Time UTC Start	Time UTC Finish	Frequency (MHz)	Mode	Power	Station	Report Sent	Report Received	Signal Sent	Signal Received

Remarks

Date	Time UTC Start	Time UTC Finish	Frequency (MHz)	Mode	Power	Station	Report Sent	Report Received	Signal Sent	Signal Received

Remarks

Date	Time UTC Start	Time UTC Finish	Frequency (MHz)	Mode	Power	Station	Report Sent	Report Received	Signal Sent	Signal Received

Remarks

Date	Time UTC Start	Time UTC Finish	Frequency (MHz)	Mode	Power	Station	Report Sent	Report Received	Signal Sent	Signal Received

Remarks

AMATEUR RADIO STATION LOG

Date	Time UTC Start	Finish	Frequency (MHz)	Mode	Power	Station	Report Sent	Received	Signal Sent	Received

Remarks

Date	Time UTC Start	Finish	Frequency (MHz)	Mode	Power	Station	Report Sent	Received	Signal Sent	Received

Remarks

Date	Time UTC Start	Finish	Frequency (MHz)	Mode	Power	Station	Report Sent	Received	Signal Sent	Received

Remarks

Date	Time UTC Start	Finish	Frequency (MHz)	Mode	Power	Station	Report Sent	Received	Signal Sent	Received

Remarks

Date	Time UTC Start	Finish	Frequency (MHz)	Mode	Power	Station	Report Sent	Received	Signal Sent	Received

Remarks

Date	Time UTC Start	Finish	Frequency (MHz)	Mode	Power	Station	Report Sent	Received	Signal Sent	Received

Remarks

Date	Time UTC Start	Finish	Frequency (MHz)	Mode	Power	Station	Report Sent	Received	Signal Sent	Received

Remarks

AMATEUR RADIO STATION LOG

Date	Time UTC Start	Time UTC Finish	Frequency (MHz)	Mode	Power	Station	Report Sent	Report Received	Signal Sent	Signal Received

Remarks

Date	Time UTC Start	Time UTC Finish	Frequency (MHz)	Mode	Power	Station	Report Sent	Report Received	Signal Sent	Signal Received

Remarks

Date	Time UTC Start	Time UTC Finish	Frequency (MHz)	Mode	Power	Station	Report Sent	Report Received	Signal Sent	Signal Received

Remarks

Date	Time UTC Start	Time UTC Finish	Frequency (MHz)	Mode	Power	Station	Report Sent	Report Received	Signal Sent	Signal Received

Remarks

Date	Time UTC Start	Time UTC Finish	Frequency (MHz)	Mode	Power	Station	Report Sent	Report Received	Signal Sent	Signal Received

Remarks

Date	Time UTC Start	Time UTC Finish	Frequency (MHz)	Mode	Power	Station	Report Sent	Report Received	Signal Sent	Signal Received

Remarks

Date	Time UTC Start	Time UTC Finish	Frequency (MHz)	Mode	Power	Station	Report Sent	Report Received	Signal Sent	Signal Received

Remarks

AMATEUR RADIO STATION LOG

Date	Time UTC Start	Finish	Frequency (MHz)	Mode	Power	Station	Report Sent	Received	Signal Sent	Received

Remarks

Date	Time UTC Start	Finish	Frequency (MHz)	Mode	Power	Station	Report Sent	Received	Signal Sent	Received

Remarks

Date	Time UTC Start	Finish	Frequency (MHz)	Mode	Power	Station	Report Sent	Received	Signal Sent	Received

Remarks

Date	Time UTC Start	Finish	Frequency (MHz)	Mode	Power	Station	Report Sent	Received	Signal Sent	Received

Remarks

Date	Time UTC Start	Finish	Frequency (MHz)	Mode	Power	Station	Report Sent	Received	Signal Sent	Received

Remarks

Date	Time UTC Start	Finish	Frequency (MHz)	Mode	Power	Station	Report Sent	Received	Signal Sent	Received

Remarks

Date	Time UTC Start	Finish	Frequency (MHz)	Mode	Power	Station	Report Sent	Received	Signal Sent	Received

Remarks

AMATEUR RADIO STATION LOG

Date	Time UTC Start	Finish	Frequency (MHz)	Mode	Power	Station	Report Sent	Received	Signal Sent	Received

Remarks

Date	Time UTC Start	Finish	Frequency (MHz)	Mode	Power	Station	Report Sent	Received	Signal Sent	Received

Remarks

Date	Time UTC Start	Finish	Frequency (MHz)	Mode	Power	Station	Report Sent	Received	Signal Sent	Received

Remarks

Date	Time UTC Start	Finish	Frequency (MHz)	Mode	Power	Station	Report Sent	Received	Signal Sent	Received

Remarks

Date	Time UTC Start	Finish	Frequency (MHz)	Mode	Power	Station	Report Sent	Received	Signal Sent	Received

Remarks

Date	Time UTC Start	Finish	Frequency (MHz)	Mode	Power	Station	Report Sent	Received	Signal Sent	Received

Remarks

Date	Time UTC Start	Finish	Frequency (MHz)	Mode	Power	Station	Report Sent	Received	Signal Sent	Received

Remarks

AMATEUR RADIO STATION LOG

Date	Time UTC Start	Finish	Frequency (MHz)	Mode	Power	Station	Report Sent	Received	Signal Sent	Received

Remarks

Date	Time UTC Start	Finish	Frequency (MHz)	Mode	Power	Station	Report Sent	Received	Signal Sent	Received

Remarks

Date	Time UTC Start	Finish	Frequency (MHz)	Mode	Power	Station	Report Sent	Received	Signal Sent	Received

Remarks

Date	Time UTC Start	Finish	Frequency (MHz)	Mode	Power	Station	Report Sent	Received	Signal Sent	Received

Remarks

Date	Time UTC Start	Finish	Frequency (MHz)	Mode	Power	Station	Report Sent	Received	Signal Sent	Received

Remarks

Date	Time UTC Start	Finish	Frequency (MHz)	Mode	Power	Station	Report Sent	Received	Signal Sent	Received

Remarks

Date	Time UTC Start	Finish	Frequency (MHz)	Mode	Power	Station	Report Sent	Received	Signal Sent	Received

Remarks

AMATEUR RADIO STATION LOG

Date	Time UTC Start	Finish	Frequency (MHz)	Mode	Power	Station	Report Sent	Received	Signal Sent	Received

Remarks

Date	Time UTC Start	Finish	Frequency (MHz)	Mode	Power	Station	Report Sent	Received	Signal Sent	Received

Remarks

Date	Time UTC Start	Finish	Frequency (MHz)	Mode	Power	Station	Report Sent	Received	Signal Sent	Received

Remarks

Date	Time UTC Start	Finish	Frequency (MHz)	Mode	Power	Station	Report Sent	Received	Signal Sent	Received

Remarks

Date	Time UTC Start	Finish	Frequency (MHz)	Mode	Power	Station	Report Sent	Received	Signal Sent	Received

Remarks

Date	Time UTC Start	Finish	Frequency (MHz)	Mode	Power	Station	Report Sent	Received	Signal Sent	Received

Remarks

Date	Time UTC Start	Finish	Frequency (MHz)	Mode	Power	Station	Report Sent	Received	Signal Sent	Received

Remarks

AMATEUR RADIO STATION LOG

Date	Time UTC Start	Finish	Frequency (MHz)	Mode	Power	Station	Report Sent	Received	Signal Sent	Received

Remarks

Date	Time UTC Start	Finish	Frequency (MHz)	Mode	Power	Station	Report Sent	Received	Signal Sent	Received

Remarks

Date	Time UTC Start	Finish	Frequency (MHz)	Mode	Power	Station	Report Sent	Received	Signal Sent	Received

Remarks

Date	Time UTC Start	Finish	Frequency (MHz)	Mode	Power	Station	Report Sent	Received	Signal Sent	Received

Remarks

Date	Time UTC Start	Finish	Frequency (MHz)	Mode	Power	Station	Report Sent	Received	Signal Sent	Received

Remarks

Date	Time UTC Start	Finish	Frequency (MHz)	Mode	Power	Station	Report Sent	Received	Signal Sent	Received

Remarks

Date	Time UTC Start	Finish	Frequency (MHz)	Mode	Power	Station	Report Sent	Received	Signal Sent	Received

Remarks

AMATEUR RADIO STATION LOG

Date	Time UTC Start	Finish	Frequency (MHz)	Mode	Power	Station	Report Sent	Received	Signal Sent	Received
Remarks										

Date	Time UTC Start	Finish	Frequency (MHz)	Mode	Power	Station	Report Sent	Received	Signal Sent	Received
Remarks										

Date	Time UTC Start	Finish	Frequency (MHz)	Mode	Power	Station	Report Sent	Received	Signal Sent	Received
Remarks										

Date	Time UTC Start	Finish	Frequency (MHz)	Mode	Power	Station	Report Sent	Received	Signal Sent	Received
Remarks										

Date	Time UTC Start	Finish	Frequency (MHz)	Mode	Power	Station	Report Sent	Received	Signal Sent	Received
Remarks										

Date	Time UTC Start	Finish	Frequency (MHz)	Mode	Power	Station	Report Sent	Received	Signal Sent	Received
Remarks										

Date	Time UTC Start	Finish	Frequency (MHz)	Mode	Power	Station	Report Sent	Received	Signal Sent	Received
Remarks										

AMATEUR RADIO STATION LOG

Date	Time UTC		Frequency (MHz)	Mode	Power	Station	Report		Signal	
	Start	Finish					Sent	Received	Sent	Received

Remarks

Date	Time UTC		Frequency (MHz)	Mode	Power	Station	Report		Signal	
	Start	Finish					Sent	Received	Sent	Received

Remarks

Date	Time UTC		Frequency (MHz)	Mode	Power	Station	Report		Signal	
	Start	Finish					Sent	Received	Sent	Received

Remarks

Date	Time UTC		Frequency (MHz)	Mode	Power	Station	Report		Signal	
	Start	Finish					Sent	Received	Sent	Received

Remarks

Date	Time UTC		Frequency (MHz)	Mode	Power	Station	Report		Signal	
	Start	Finish					Sent	Received	Sent	Received

Remarks

Date	Time UTC		Frequency (MHz)	Mode	Power	Station	Report		Signal	
	Start	Finish					Sent	Received	Sent	Received

Remarks

Date	Time UTC		Frequency (MHz)	Mode	Power	Station	Report		Signal	
	Start	Finish					Sent	Received	Sent	Received

Remarks

AMATEUR RADIO STATION LOG

Date	Time UTC Start	Finish	Frequency (MHz)	Mode	Power	Station	Report Sent	Received	Signal Sent	Received

Remarks

Date	Time UTC Start	Finish	Frequency (MHz)	Mode	Power	Station	Report Sent	Received	Signal Sent	Received

Remarks

Date	Time UTC Start	Finish	Frequency (MHz)	Mode	Power	Station	Report Sent	Received	Signal Sent	Received

Remarks

Date	Time UTC Start	Finish	Frequency (MHz)	Mode	Power	Station	Report Sent	Received	Signal Sent	Received

Remarks

Date	Time UTC Start	Finish	Frequency (MHz)	Mode	Power	Station	Report Sent	Received	Signal Sent	Received

Remarks

Date	Time UTC Start	Finish	Frequency (MHz)	Mode	Power	Station	Report Sent	Received	Signal Sent	Received

Remarks

Date	Time UTC Start	Finish	Frequency (MHz)	Mode	Power	Station	Report Sent	Received	Signal Sent	Received

Remarks

AMATEUR RADIO STATION LOG

Date	Time UTC Start	Finish	Frequency (MHz)	Mode	Power	Station	Report Sent	Received	Signal Sent	Received

Remarks

Date	Time UTC Start	Finish	Frequency (MHz)	Mode	Power	Station	Report Sent	Received	Signal Sent	Received

Remarks

Date	Time UTC Start	Finish	Frequency (MHz)	Mode	Power	Station	Report Sent	Received	Signal Sent	Received

Remarks

Date	Time UTC Start	Finish	Frequency (MHz)	Mode	Power	Station	Report Sent	Received	Signal Sent	Received

Remarks

Date	Time UTC Start	Finish	Frequency (MHz)	Mode	Power	Station	Report Sent	Received	Signal Sent	Received

Remarks

Date	Time UTC Start	Finish	Frequency (MHz)	Mode	Power	Station	Report Sent	Received	Signal Sent	Received

Remarks

Date	Time UTC Start	Finish	Frequency (MHz)	Mode	Power	Station	Report Sent	Received	Signal Sent	Received

Remarks

AMATEUR RADIO STATION LOG

Date	Time UTC Start	Finish	Frequency (MHz)	Mode	Power	Station	Report Sent	Received	Signal Sent	Received

Remarks

Date	Time UTC Start	Finish	Frequency (MHz)	Mode	Power	Station	Report Sent	Received	Signal Sent	Received

Remarks

Date	Time UTC Start	Finish	Frequency (MHz)	Mode	Power	Station	Report Sent	Received	Signal Sent	Received

Remarks

Date	Time UTC Start	Finish	Frequency (MHz)	Mode	Power	Station	Report Sent	Received	Signal Sent	Received

Remarks

Date	Time UTC Start	Finish	Frequency (MHz)	Mode	Power	Station	Report Sent	Received	Signal Sent	Received

Remarks

Date	Time UTC Start	Finish	Frequency (MHz)	Mode	Power	Station	Report Sent	Received	Signal Sent	Received

Remarks

Date	Time UTC Start	Finish	Frequency (MHz)	Mode	Power	Station	Report Sent	Received	Signal Sent	Received

Remarks

AMATEUR RADIO STATION LOG

Date	Time UTC		Frequency (MHz)	Mode	Power	Station	Report		Signal	
	Start	Finish					Sent	Received	Sent	Received

Remarks

Date	Time UTC		Frequency (MHz)	Mode	Power	Station	Report		Signal	
	Start	Finish					Sent	Received	Sent	Received

Remarks

Date	Time UTC		Frequency (MHz)	Mode	Power	Station	Report		Signal	
	Start	Finish					Sent	Received	Sent	Received

Remarks

Date	Time UTC		Frequency (MHz)	Mode	Power	Station	Report		Signal	
	Start	Finish					Sent	Received	Sent	Received

Remarks

Date	Time UTC		Frequency (MHz)	Mode	Power	Station	Report		Signal	
	Start	Finish					Sent	Received	Sent	Received

Remarks

Date	Time UTC		Frequency (MHz)	Mode	Power	Station	Report		Signal	
	Start	Finish					Sent	Received	Sent	Received

Remarks

Date	Time UTC		Frequency (MHz)	Mode	Power	Station	Report		Signal	
	Start	Finish					Sent	Received	Sent	Received

Remarks

AMATEUR RADIO STATION LOG

Date	Time UTC Start	Finish	Frequency (MHz)	Mode	Power	Station	Report Sent	Received	Signal Sent	Received

Remarks

Date	Time UTC Start	Finish	Frequency (MHz)	Mode	Power	Station	Report Sent	Received	Signal Sent	Received

Remarks

Date	Time UTC Start	Finish	Frequency (MHz)	Mode	Power	Station	Report Sent	Received	Signal Sent	Received

Remarks

Date	Time UTC Start	Finish	Frequency (MHz)	Mode	Power	Station	Report Sent	Received	Signal Sent	Received

Remarks

Date	Time UTC Start	Finish	Frequency (MHz)	Mode	Power	Station	Report Sent	Received	Signal Sent	Received

Remarks

Date	Time UTC Start	Finish	Frequency (MHz)	Mode	Power	Station	Report Sent	Received	Signal Sent	Received

Remarks

Date	Time UTC Start	Finish	Frequency (MHz)	Mode	Power	Station	Report Sent	Received	Signal Sent	Received

Remarks

AMATEUR RADIO STATION LOG

Date	Time UTC Start	Finish	Frequency (MHz)	Mode	Power	Station	Report Sent	Received	Signal Sent	Received
Remarks										

Date	Time UTC Start	Finish	Frequency (MHz)	Mode	Power	Station	Report Sent	Received	Signal Sent	Received
Remarks										

Date	Time UTC Start	Finish	Frequency (MHz)	Mode	Power	Station	Report Sent	Received	Signal Sent	Received
Remarks										

Date	Time UTC Start	Finish	Frequency (MHz)	Mode	Power	Station	Report Sent	Received	Signal Sent	Received
Remarks										

Date	Time UTC Start	Finish	Frequency (MHz)	Mode	Power	Station	Report Sent	Received	Signal Sent	Received
Remarks										

Date	Time UTC Start	Finish	Frequency (MHz)	Mode	Power	Station	Report Sent	Received	Signal Sent	Received
Remarks										

Date	Time UTC Start	Finish	Frequency (MHz)	Mode	Power	Station	Report Sent	Received	Signal Sent	Received
Remarks										

AMATEUR RADIO STATION LOG

Date	Time UTC Start	Finish	Frequency (MHz)	Mode	Power	Station	Report Sent	Received	Signal Sent	Received

Remarks

Date	Time UTC Start	Finish	Frequency (MHz)	Mode	Power	Station	Report Sent	Received	Signal Sent	Received

Remarks

Date	Time UTC Start	Finish	Frequency (MHz)	Mode	Power	Station	Report Sent	Received	Signal Sent	Received

Remarks

Date	Time UTC Start	Finish	Frequency (MHz)	Mode	Power	Station	Report Sent	Received	Signal Sent	Received

Remarks

Date	Time UTC Start	Finish	Frequency (MHz)	Mode	Power	Station	Report Sent	Received	Signal Sent	Received

Remarks

Date	Time UTC Start	Finish	Frequency (MHz)	Mode	Power	Station	Report Sent	Received	Signal Sent	Received

Remarks

Date	Time UTC Start	Finish	Frequency (MHz)	Mode	Power	Station	Report Sent	Received	Signal Sent	Received

Remarks

AMATEUR RADIO STATION LOG

Date	Time UTC Start	Finish	Frequency (MHz)	Mode	Power	Station	Report Sent	Received	Signal Sent	Received

Remarks

Date	Time UTC Start	Finish	Frequency (MHz)	Mode	Power	Station	Report Sent	Received	Signal Sent	Received

Remarks

Date	Time UTC Start	Finish	Frequency (MHz)	Mode	Power	Station	Report Sent	Received	Signal Sent	Received

Remarks

Date	Time UTC Start	Finish	Frequency (MHz)	Mode	Power	Station	Report Sent	Received	Signal Sent	Received

Remarks

Date	Time UTC Start	Finish	Frequency (MHz)	Mode	Power	Station	Report Sent	Received	Signal Sent	Received

Remarks

Date	Time UTC Start	Finish	Frequency (MHz)	Mode	Power	Station	Report Sent	Received	Signal Sent	Received

Remarks

Date	Time UTC Start	Finish	Frequency (MHz)	Mode	Power	Station	Report Sent	Received	Signal Sent	Received

Remarks

AMATEUR RADIO STATION LOG

Date	Time UTC Start	Finish	Frequency (MHz)	Mode	Power	Station	Report Sent	Received	Signal Sent	Received
Remarks										

Date	Time UTC Start	Finish	Frequency (MHz)	Mode	Power	Station	Report Sent	Received	Signal Sent	Received
Remarks										

Date	Time UTC Start	Finish	Frequency (MHz)	Mode	Power	Station	Report Sent	Received	Signal Sent	Received
Remarks										

Date	Time UTC Start	Finish	Frequency (MHz)	Mode	Power	Station	Report Sent	Received	Signal Sent	Received
Remarks										

Date	Time UTC Start	Finish	Frequency (MHz)	Mode	Power	Station	Report Sent	Received	Signal Sent	Received
Remarks										

Date	Time UTC Start	Finish	Frequency (MHz)	Mode	Power	Station	Report Sent	Received	Signal Sent	Received
Remarks										

Date	Time UTC Start	Finish	Frequency (MHz)	Mode	Power	Station	Report Sent	Received	Signal Sent	Received
Remarks										

AMATEUR RADIO STATION LOG

Date	Time UTC Start	Finish	Frequency (MHz)	Mode	Power	Station	Report Sent	Received	Signal Sent	Received

Remarks

Date	Time UTC Start	Finish	Frequency (MHz)	Mode	Power	Station	Report Sent	Received	Signal Sent	Received

Remarks

Date	Time UTC Start	Finish	Frequency (MHz)	Mode	Power	Station	Report Sent	Received	Signal Sent	Received

Remarks

Date	Time UTC Start	Finish	Frequency (MHz)	Mode	Power	Station	Report Sent	Received	Signal Sent	Received

Remarks

Date	Time UTC Start	Finish	Frequency (MHz)	Mode	Power	Station	Report Sent	Received	Signal Sent	Received

Remarks

Date	Time UTC Start	Finish	Frequency (MHz)	Mode	Power	Station	Report Sent	Received	Signal Sent	Received

Remarks

Date	Time UTC Start	Finish	Frequency (MHz)	Mode	Power	Station	Report Sent	Received	Signal Sent	Received

Remarks

AMATEUR RADIO STATION LOG

Date	Time UTC Start	Time UTC Finish	Frequency (MHz)	Mode	Power	Station	Report Sent	Report Received	Signal Sent	Signal Received

Remarks

Date	Time UTC Start	Time UTC Finish	Frequency (MHz)	Mode	Power	Station	Report Sent	Report Received	Signal Sent	Signal Received

Remarks

Date	Time UTC Start	Time UTC Finish	Frequency (MHz)	Mode	Power	Station	Report Sent	Report Received	Signal Sent	Signal Received

Remarks

Date	Time UTC Start	Time UTC Finish	Frequency (MHz)	Mode	Power	Station	Report Sent	Report Received	Signal Sent	Signal Received

Remarks

Date	Time UTC Start	Time UTC Finish	Frequency (MHz)	Mode	Power	Station	Report Sent	Report Received	Signal Sent	Signal Received

Remarks

Date	Time UTC Start	Time UTC Finish	Frequency (MHz)	Mode	Power	Station	Report Sent	Report Received	Signal Sent	Signal Received

Remarks

Date	Time UTC Start	Time UTC Finish	Frequency (MHz)	Mode	Power	Station	Report Sent	Report Received	Signal Sent	Signal Received

Remarks

AMATEUR RADIO STATION LOG

Date	Time UTC Start	Finish	Frequency (MHz)	Mode	Power	Station	Report Sent	Received	Signal Sent	Received
Remarks										

Date	Time UTC Start	Finish	Frequency (MHz)	Mode	Power	Station	Report Sent	Received	Signal Sent	Received
Remarks										

Date	Time UTC Start	Finish	Frequency (MHz)	Mode	Power	Station	Report Sent	Received	Signal Sent	Received
Remarks										

Date	Time UTC Start	Finish	Frequency (MHz)	Mode	Power	Station	Report Sent	Received	Signal Sent	Received
Remarks										

Date	Time UTC Start	Finish	Frequency (MHz)	Mode	Power	Station	Report Sent	Received	Signal Sent	Received
Remarks										

Date	Time UTC Start	Finish	Frequency (MHz)	Mode	Power	Station	Report Sent	Received	Signal Sent	Received
Remarks										

Date	Time UTC Start	Finish	Frequency (MHz)	Mode	Power	Station	Report Sent	Received	Signal Sent	Received
Remarks										

AMATEUR RADIO STATION LOG

Date	Time UTC Start	Time UTC Finish	Frequency (MHz)	Mode	Power	Station	Report Sent	Report Received	Signal Sent	Signal Received

Remarks

Date	Time UTC Start	Time UTC Finish	Frequency (MHz)	Mode	Power	Station	Report Sent	Report Received	Signal Sent	Signal Received

Remarks

Date	Time UTC Start	Time UTC Finish	Frequency (MHz)	Mode	Power	Station	Report Sent	Report Received	Signal Sent	Signal Received

Remarks

Date	Time UTC Start	Time UTC Finish	Frequency (MHz)	Mode	Power	Station	Report Sent	Report Received	Signal Sent	Signal Received

Remarks

Date	Time UTC Start	Time UTC Finish	Frequency (MHz)	Mode	Power	Station	Report Sent	Report Received	Signal Sent	Signal Received

Remarks

Date	Time UTC Start	Time UTC Finish	Frequency (MHz)	Mode	Power	Station	Report Sent	Report Received	Signal Sent	Signal Received

Remarks

Date	Time UTC Start	Time UTC Finish	Frequency (MHz)	Mode	Power	Station	Report Sent	Report Received	Signal Sent	Signal Received

Remarks

AMATEUR RADIO STATION LOG

Date	Time UTC Start	Finish	Frequency (MHz)	Mode	Power	Station	Report Sent	Received	Signal Sent	Received

Remarks

Date	Time UTC Start	Finish	Frequency (MHz)	Mode	Power	Station	Report Sent	Received	Signal Sent	Received

Remarks

Date	Time UTC Start	Finish	Frequency (MHz)	Mode	Power	Station	Report Sent	Received	Signal Sent	Received

Remarks

Date	Time UTC Start	Finish	Frequency (MHz)	Mode	Power	Station	Report Sent	Received	Signal Sent	Received

Remarks

Date	Time UTC Start	Finish	Frequency (MHz)	Mode	Power	Station	Report Sent	Received	Signal Sent	Received

Remarks

Date	Time UTC Start	Finish	Frequency (MHz)	Mode	Power	Station	Report Sent	Received	Signal Sent	Received

Remarks

Date	Time UTC Start	Finish	Frequency (MHz)	Mode	Power	Station	Report Sent	Received	Signal Sent	Received

Remarks

AMATEUR RADIO STATION LOG

Date	Time UTC Start	Time UTC Finish	Frequency (MHz)	Mode	Power	Station	Report Sent	Report Received	Signal Sent	Signal Received

Remarks

Date	Time UTC Start	Time UTC Finish	Frequency (MHz)	Mode	Power	Station	Report Sent	Report Received	Signal Sent	Signal Received

Remarks

Date	Time UTC Start	Time UTC Finish	Frequency (MHz)	Mode	Power	Station	Report Sent	Report Received	Signal Sent	Signal Received

Remarks

Date	Time UTC Start	Time UTC Finish	Frequency (MHz)	Mode	Power	Station	Report Sent	Report Received	Signal Sent	Signal Received

Remarks

Date	Time UTC Start	Time UTC Finish	Frequency (MHz)	Mode	Power	Station	Report Sent	Report Received	Signal Sent	Signal Received

Remarks

Date	Time UTC Start	Time UTC Finish	Frequency (MHz)	Mode	Power	Station	Report Sent	Report Received	Signal Sent	Signal Received

Remarks

Date	Time UTC Start	Time UTC Finish	Frequency (MHz)	Mode	Power	Station	Report Sent	Report Received	Signal Sent	Signal Received

Remarks

AMATEUR RADIO STATION LOG

Date	Time UTC Start	Finish	Frequency (MHz)	Mode	Power	Station	Report Sent	Received	Signal Sent	Received

Remarks

Date	Time UTC Start	Finish	Frequency (MHz)	Mode	Power	Station	Report Sent	Received	Signal Sent	Received

Remarks

Date	Time UTC Start	Finish	Frequency (MHz)	Mode	Power	Station	Report Sent	Received	Signal Sent	Received

Remarks

Date	Time UTC Start	Finish	Frequency (MHz)	Mode	Power	Station	Report Sent	Received	Signal Sent	Received

Remarks

Date	Time UTC Start	Finish	Frequency (MHz)	Mode	Power	Station	Report Sent	Received	Signal Sent	Received

Remarks

Date	Time UTC Start	Finish	Frequency (MHz)	Mode	Power	Station	Report Sent	Received	Signal Sent	Received

Remarks

Date	Time UTC Start	Finish	Frequency (MHz)	Mode	Power	Station	Report Sent	Received	Signal Sent	Received

Remarks

AMATEUR RADIO STATION LOG

Date	Time UTC		Frequency (MHz)	Mode	Power	Station	Report		Signal	
	Start	Finish					Sent	Received	Sent	Received

Remarks

Date	Time UTC		Frequency (MHz)	Mode	Power	Station	Report		Signal	
	Start	Finish					Sent	Received	Sent	Received

Remarks

Date	Time UTC		Frequency (MHz)	Mode	Power	Station	Report		Signal	
	Start	Finish					Sent	Received	Sent	Received

Remarks

Date	Time UTC		Frequency (MHz)	Mode	Power	Station	Report		Signal	
	Start	Finish					Sent	Received	Sent	Received

Remarks

Date	Time UTC		Frequency (MHz)	Mode	Power	Station	Report		Signal	
	Start	Finish					Sent	Received	Sent	Received

Remarks

Date	Time UTC		Frequency (MHz)	Mode	Power	Station	Report		Signal	
	Start	Finish					Sent	Received	Sent	Received

Remarks

Date	Time UTC		Frequency (MHz)	Mode	Power	Station	Report		Signal	
	Start	Finish					Sent	Received	Sent	Received

Remarks

AMATEUR RADIO STATION LOG

Date	Time UTC		Frequency (MHz)	Mode	Power	Station	Report		Signal	
	Start	Finish					Sent	Received	Sent	Received

Remarks

Date	Time UTC		Frequency (MHz)	Mode	Power	Station	Report		Signal	
	Start	Finish					Sent	Received	Sent	Received

Remarks

Date	Time UTC		Frequency (MHz)	Mode	Power	Station	Report		Signal	
	Start	Finish					Sent	Received	Sent	Received

Remarks

Date	Time UTC		Frequency (MHz)	Mode	Power	Station	Report		Signal	
	Start	Finish					Sent	Received	Sent	Received

Remarks

Date	Time UTC		Frequency (MHz)	Mode	Power	Station	Report		Signal	
	Start	Finish					Sent	Received	Sent	Received

Remarks

Date	Time UTC		Frequency (MHz)	Mode	Power	Station	Report		Signal	
	Start	Finish					Sent	Received	Sent	Received

Remarks

Date	Time UTC		Frequency (MHz)	Mode	Power	Station	Report		Signal	
	Start	Finish					Sent	Received	Sent	Received

Remarks

AMATEUR RADIO STATION LOG

Date	Time UTC Start	Time UTC Finish	Frequency (MHz)	Mode	Power	Station	Report Sent	Report Received	Signal Sent	Signal Received

Remarks

Date	Time UTC Start	Time UTC Finish	Frequency (MHz)	Mode	Power	Station	Report Sent	Report Received	Signal Sent	Signal Received

Remarks

Date	Time UTC Start	Time UTC Finish	Frequency (MHz)	Mode	Power	Station	Report Sent	Report Received	Signal Sent	Signal Received

Remarks

Date	Time UTC Start	Time UTC Finish	Frequency (MHz)	Mode	Power	Station	Report Sent	Report Received	Signal Sent	Signal Received

Remarks

Date	Time UTC Start	Time UTC Finish	Frequency (MHz)	Mode	Power	Station	Report Sent	Report Received	Signal Sent	Signal Received

Remarks

Date	Time UTC Start	Time UTC Finish	Frequency (MHz)	Mode	Power	Station	Report Sent	Report Received	Signal Sent	Signal Received

Remarks

Date	Time UTC Start	Time UTC Finish	Frequency (MHz)	Mode	Power	Station	Report Sent	Report Received	Signal Sent	Signal Received

Remarks

AMATEUR RADIO STATION LOG

Date	Time UTC Start	Finish	Frequency (MHz)	Mode	Power	Station	Report Sent	Received	Signal Sent	Received

Remarks

Date	Time UTC Start	Finish	Frequency (MHz)	Mode	Power	Station	Report Sent	Received	Signal Sent	Received

Remarks

Date	Time UTC Start	Finish	Frequency (MHz)	Mode	Power	Station	Report Sent	Received	Signal Sent	Received

Remarks

Date	Time UTC Start	Finish	Frequency (MHz)	Mode	Power	Station	Report Sent	Received	Signal Sent	Received

Remarks

Date	Time UTC Start	Finish	Frequency (MHz)	Mode	Power	Station	Report Sent	Received	Signal Sent	Received

Remarks

Date	Time UTC Start	Finish	Frequency (MHz)	Mode	Power	Station	Report Sent	Received	Signal Sent	Received

Remarks

Date	Time UTC Start	Finish	Frequency (MHz)	Mode	Power	Station	Report Sent	Received	Signal Sent	Received

Remarks

AMATEUR RADIO STATION LOG

Date	Time UTC Start	Finish	Frequency (MHz)	Mode	Power	Station	Report Sent	Received	Signal Sent	Received
Remarks										

Date	Time UTC Start	Finish	Frequency (MHz)	Mode	Power	Station	Report Sent	Received	Signal Sent	Received
Remarks										

Date	Time UTC Start	Finish	Frequency (MHz)	Mode	Power	Station	Report Sent	Received	Signal Sent	Received
Remarks										

Date	Time UTC Start	Finish	Frequency (MHz)	Mode	Power	Station	Report Sent	Received	Signal Sent	Received
Remarks										

Date	Time UTC Start	Finish	Frequency (MHz)	Mode	Power	Station	Report Sent	Received	Signal Sent	Received
Remarks										

Date	Time UTC Start	Finish	Frequency (MHz)	Mode	Power	Station	Report Sent	Received	Signal Sent	Received
Remarks										

Date	Time UTC Start	Finish	Frequency (MHz)	Mode	Power	Station	Report Sent	Received	Signal Sent	Received
Remarks										

AMATEUR RADIO STATION LOG

Date	Time UTC Start	Finish	Frequency (MHz)	Mode	Power	Station	Report Sent	Received	Signal Sent	Received

Remarks

Date	Time UTC Start	Finish	Frequency (MHz)	Mode	Power	Station	Report Sent	Received	Signal Sent	Received

Remarks

Date	Time UTC Start	Finish	Frequency (MHz)	Mode	Power	Station	Report Sent	Received	Signal Sent	Received

Remarks

Date	Time UTC Start	Finish	Frequency (MHz)	Mode	Power	Station	Report Sent	Received	Signal Sent	Received

Remarks

Date	Time UTC Start	Finish	Frequency (MHz)	Mode	Power	Station	Report Sent	Received	Signal Sent	Received

Remarks

Date	Time UTC Start	Finish	Frequency (MHz)	Mode	Power	Station	Report Sent	Received	Signal Sent	Received

Remarks

Date	Time UTC Start	Finish	Frequency (MHz)	Mode	Power	Station	Report Sent	Received	Signal Sent	Received

Remarks

AMATEUR RADIO STATION LOG

Date	Time UTC Start	Finish	Frequency (MHz)	Mode	Power	Station	Report Sent	Received	Signal Sent	Received

Remarks

Date	Time UTC Start	Finish	Frequency (MHz)	Mode	Power	Station	Report Sent	Received	Signal Sent	Received

Remarks

Date	Time UTC Start	Finish	Frequency (MHz)	Mode	Power	Station	Report Sent	Received	Signal Sent	Received

Remarks

Date	Time UTC Start	Finish	Frequency (MHz)	Mode	Power	Station	Report Sent	Received	Signal Sent	Received

Remarks

Date	Time UTC Start	Finish	Frequency (MHz)	Mode	Power	Station	Report Sent	Received	Signal Sent	Received

Remarks

Date	Time UTC Start	Finish	Frequency (MHz)	Mode	Power	Station	Report Sent	Received	Signal Sent	Received

Remarks

Date	Time UTC Start	Finish	Frequency (MHz)	Mode	Power	Station	Report Sent	Received	Signal Sent	Received

Remarks

AMATEUR RADIO STATION LOG

Date	Time UTC Start	Finish	Frequency (MHz)	Mode	Power	Station	Report Sent	Received	Signal Sent	Received

Remarks

Date	Time UTC Start	Finish	Frequency (MHz)	Mode	Power	Station	Report Sent	Received	Signal Sent	Received

Remarks

Date	Time UTC Start	Finish	Frequency (MHz)	Mode	Power	Station	Report Sent	Received	Signal Sent	Received

Remarks

Date	Time UTC Start	Finish	Frequency (MHz)	Mode	Power	Station	Report Sent	Received	Signal Sent	Received

Remarks

Date	Time UTC Start	Finish	Frequency (MHz)	Mode	Power	Station	Report Sent	Received	Signal Sent	Received

Remarks

Date	Time UTC Start	Finish	Frequency (MHz)	Mode	Power	Station	Report Sent	Received	Signal Sent	Received

Remarks

Date	Time UTC Start	Finish	Frequency (MHz)	Mode	Power	Station	Report Sent	Received	Signal Sent	Received

Remarks

AMATEUR RADIO STATION LOG

Date	Time UTC Start	Time UTC Finish	Frequency (MHz)	Mode	Power	Station	Report Sent	Report Received	Signal Sent	Signal Received

Remarks

Date	Time UTC Start	Time UTC Finish	Frequency (MHz)	Mode	Power	Station	Report Sent	Report Received	Signal Sent	Signal Received

Remarks

Date	Time UTC Start	Time UTC Finish	Frequency (MHz)	Mode	Power	Station	Report Sent	Report Received	Signal Sent	Signal Received

Remarks

Date	Time UTC Start	Time UTC Finish	Frequency (MHz)	Mode	Power	Station	Report Sent	Report Received	Signal Sent	Signal Received

Remarks

Date	Time UTC Start	Time UTC Finish	Frequency (MHz)	Mode	Power	Station	Report Sent	Report Received	Signal Sent	Signal Received

Remarks

Date	Time UTC Start	Time UTC Finish	Frequency (MHz)	Mode	Power	Station	Report Sent	Report Received	Signal Sent	Signal Received

Remarks

Date	Time UTC Start	Time UTC Finish	Frequency (MHz)	Mode	Power	Station	Report Sent	Report Received	Signal Sent	Signal Received

Remarks

AMATEUR RADIO STATION LOG

Date	Time UTC Start	Finish	Frequency (MHz)	Mode	Power	Station	Report Sent	Received	Signal Sent	Received

Remarks

Date	Time UTC Start	Finish	Frequency (MHz)	Mode	Power	Station	Report Sent	Received	Signal Sent	Received

Remarks

Date	Time UTC Start	Finish	Frequency (MHz)	Mode	Power	Station	Report Sent	Received	Signal Sent	Received

Remarks

Date	Time UTC Start	Finish	Frequency (MHz)	Mode	Power	Station	Report Sent	Received	Signal Sent	Received

Remarks

Date	Time UTC Start	Finish	Frequency (MHz)	Mode	Power	Station	Report Sent	Received	Signal Sent	Received

Remarks

Date	Time UTC Start	Finish	Frequency (MHz)	Mode	Power	Station	Report Sent	Received	Signal Sent	Received

Remarks

Date	Time UTC Start	Finish	Frequency (MHz)	Mode	Power	Station	Report Sent	Received	Signal Sent	Received

Remarks

AMATEUR RADIO STATION LOG

Date	Time UTC Start	Time UTC Finish	Frequency (MHz)	Mode	Power	Station	Report Sent	Report Received	Signal Sent	Signal Received

Remarks

Date	Time UTC Start	Time UTC Finish	Frequency (MHz)	Mode	Power	Station	Report Sent	Report Received	Signal Sent	Signal Received

Remarks

Date	Time UTC Start	Time UTC Finish	Frequency (MHz)	Mode	Power	Station	Report Sent	Report Received	Signal Sent	Signal Received

Remarks

Date	Time UTC Start	Time UTC Finish	Frequency (MHz)	Mode	Power	Station	Report Sent	Report Received	Signal Sent	Signal Received

Remarks

Date	Time UTC Start	Time UTC Finish	Frequency (MHz)	Mode	Power	Station	Report Sent	Report Received	Signal Sent	Signal Received

Remarks

Date	Time UTC Start	Time UTC Finish	Frequency (MHz)	Mode	Power	Station	Report Sent	Report Received	Signal Sent	Signal Received

Remarks

Date	Time UTC Start	Time UTC Finish	Frequency (MHz)	Mode	Power	Station	Report Sent	Report Received	Signal Sent	Signal Received

Remarks

AMATEUR RADIO STATION LOG

Date	Time UTC		Frequency (MHz)	Mode	Power	Station	Report		Signal	
	Start	Finish					Sent	Received	Sent	Received

Remarks

Date	Time UTC		Frequency (MHz)	Mode	Power	Station	Report		Signal	
	Start	Finish					Sent	Received	Sent	Received

Remarks

Date	Time UTC		Frequency (MHz)	Mode	Power	Station	Report		Signal	
	Start	Finish					Sent	Received	Sent	Received

Remarks

Date	Time UTC		Frequency (MHz)	Mode	Power	Station	Report		Signal	
	Start	Finish					Sent	Received	Sent	Received

Remarks

Date	Time UTC		Frequency (MHz)	Mode	Power	Station	Report		Signal	
	Start	Finish					Sent	Received	Sent	Received

Remarks

Date	Time UTC		Frequency (MHz)	Mode	Power	Station	Report		Signal	
	Start	Finish					Sent	Received	Sent	Received

Remarks

Date	Time UTC		Frequency (MHz)	Mode	Power	Station	Report		Signal	
	Start	Finish					Sent	Received	Sent	Received

Remarks

AMATEUR RADIO STATION LOG

Date	Time UTC Start	Finish	Frequency (MHz)	Mode	Power	Station	Report Sent	Received	Signal Sent	Received
Remarks										

Date	Time UTC Start	Finish	Frequency (MHz)	Mode	Power	Station	Report Sent	Received	Signal Sent	Received
Remarks										

Date	Time UTC Start	Finish	Frequency (MHz)	Mode	Power	Station	Report Sent	Received	Signal Sent	Received
Remarks										

Date	Time UTC Start	Finish	Frequency (MHz)	Mode	Power	Station	Report Sent	Received	Signal Sent	Received
Remarks										

Date	Time UTC Start	Finish	Frequency (MHz)	Mode	Power	Station	Report Sent	Received	Signal Sent	Received
Remarks										

Date	Time UTC Start	Finish	Frequency (MHz)	Mode	Power	Station	Report Sent	Received	Signal Sent	Received
Remarks										

Date	Time UTC Start	Finish	Frequency (MHz)	Mode	Power	Station	Report Sent	Received	Signal Sent	Received
Remarks										

AMATEUR RADIO STATION LOG

Date	Time UTC Start	Time UTC Finish	Frequency (MHz)	Mode	Power	Station	Report Sent	Report Received	Signal Sent	Signal Received

Remarks

Date	Time UTC Start	Time UTC Finish	Frequency (MHz)	Mode	Power	Station	Report Sent	Report Received	Signal Sent	Signal Received

Remarks

Date	Time UTC Start	Time UTC Finish	Frequency (MHz)	Mode	Power	Station	Report Sent	Report Received	Signal Sent	Signal Received

Remarks

Date	Time UTC Start	Time UTC Finish	Frequency (MHz)	Mode	Power	Station	Report Sent	Report Received	Signal Sent	Signal Received

Remarks

Date	Time UTC Start	Time UTC Finish	Frequency (MHz)	Mode	Power	Station	Report Sent	Report Received	Signal Sent	Signal Received

Remarks

Date	Time UTC Start	Time UTC Finish	Frequency (MHz)	Mode	Power	Station	Report Sent	Report Received	Signal Sent	Signal Received

Remarks

Date	Time UTC Start	Time UTC Finish	Frequency (MHz)	Mode	Power	Station	Report Sent	Report Received	Signal Sent	Signal Received

Remarks

AMATEUR RADIO STATION LOG

Date	Time UTC Start	Time UTC Finish	Frequency (MHz)	Mode	Power	Station	Report Sent	Report Received	Signal Sent	Signal Received

Remarks

Date	Time UTC Start	Time UTC Finish	Frequency (MHz)	Mode	Power	Station	Report Sent	Report Received	Signal Sent	Signal Received

Remarks

Date	Time UTC Start	Time UTC Finish	Frequency (MHz)	Mode	Power	Station	Report Sent	Report Received	Signal Sent	Signal Received

Remarks

Date	Time UTC Start	Time UTC Finish	Frequency (MHz)	Mode	Power	Station	Report Sent	Report Received	Signal Sent	Signal Received

Remarks

Date	Time UTC Start	Time UTC Finish	Frequency (MHz)	Mode	Power	Station	Report Sent	Report Received	Signal Sent	Signal Received

Remarks

Date	Time UTC Start	Time UTC Finish	Frequency (MHz)	Mode	Power	Station	Report Sent	Report Received	Signal Sent	Signal Received

Remarks

Date	Time UTC Start	Time UTC Finish	Frequency (MHz)	Mode	Power	Station	Report Sent	Report Received	Signal Sent	Signal Received

Remarks

AMATEUR RADIO STATION LOG

Date	Time UTC Start	Finish	Frequency (MHz)	Mode	Power	Station	Report Sent	Received	Signal Sent	Received

Remarks

Date	Time UTC Start	Finish	Frequency (MHz)	Mode	Power	Station	Report Sent	Received	Signal Sent	Received

Remarks

Date	Time UTC Start	Finish	Frequency (MHz)	Mode	Power	Station	Report Sent	Received	Signal Sent	Received

Remarks

Date	Time UTC Start	Finish	Frequency (MHz)	Mode	Power	Station	Report Sent	Received	Signal Sent	Received

Remarks

Date	Time UTC Start	Finish	Frequency (MHz)	Mode	Power	Station	Report Sent	Received	Signal Sent	Received

Remarks

Date	Time UTC Start	Finish	Frequency (MHz)	Mode	Power	Station	Report Sent	Received	Signal Sent	Received

Remarks

Date	Time UTC Start	Finish	Frequency (MHz)	Mode	Power	Station	Report Sent	Received	Signal Sent	Received

Remarks

AMATEUR RADIO STATION LOG

Date	Time UTC Start	Finish	Frequency (MHz)	Mode	Power	Station	Report Sent	Received	Signal Sent	Received
Remarks										

Date	Time UTC Start	Finish	Frequency (MHz)	Mode	Power	Station	Report Sent	Received	Signal Sent	Received
Remarks										

Date	Time UTC Start	Finish	Frequency (MHz)	Mode	Power	Station	Report Sent	Received	Signal Sent	Received
Remarks										

Date	Time UTC Start	Finish	Frequency (MHz)	Mode	Power	Station	Report Sent	Received	Signal Sent	Received
Remarks										

Date	Time UTC Start	Finish	Frequency (MHz)	Mode	Power	Station	Report Sent	Received	Signal Sent	Received
Remarks										

Date	Time UTC Start	Finish	Frequency (MHz)	Mode	Power	Station	Report Sent	Received	Signal Sent	Received
Remarks										

Date	Time UTC Start	Finish	Frequency (MHz)	Mode	Power	Station	Report Sent	Received	Signal Sent	Received
Remarks										

AMATEUR RADIO STATION LOG

Date	Time UTC Start	Time UTC Finish	Frequency (MHz)	Mode	Power	Station	Report Sent	Report Received	Signal Sent	Signal Received
Remarks										

Date	Time UTC Start	Time UTC Finish	Frequency (MHz)	Mode	Power	Station	Report Sent	Report Received	Signal Sent	Signal Received
Remarks										

Date	Time UTC Start	Time UTC Finish	Frequency (MHz)	Mode	Power	Station	Report Sent	Report Received	Signal Sent	Signal Received
Remarks										

Date	Time UTC Start	Time UTC Finish	Frequency (MHz)	Mode	Power	Station	Report Sent	Report Received	Signal Sent	Signal Received
Remarks										

Date	Time UTC Start	Time UTC Finish	Frequency (MHz)	Mode	Power	Station	Report Sent	Report Received	Signal Sent	Signal Received
Remarks										

Date	Time UTC Start	Time UTC Finish	Frequency (MHz)	Mode	Power	Station	Report Sent	Report Received	Signal Sent	Signal Received
Remarks										

Date	Time UTC Start	Time UTC Finish	Frequency (MHz)	Mode	Power	Station	Report Sent	Report Received	Signal Sent	Signal Received
Remarks										

AMATEUR RADIO STATION LOG

Date	Time UTC Start	Finish	Frequency (MHz)	Mode	Power	Station	Report Sent	Received	Signal Sent	Received

Remarks

Date	Time UTC Start	Finish	Frequency (MHz)	Mode	Power	Station	Report Sent	Received	Signal Sent	Received

Remarks

Date	Time UTC Start	Finish	Frequency (MHz)	Mode	Power	Station	Report Sent	Received	Signal Sent	Received

Remarks

Date	Time UTC Start	Finish	Frequency (MHz)	Mode	Power	Station	Report Sent	Received	Signal Sent	Received

Remarks

Date	Time UTC Start	Finish	Frequency (MHz)	Mode	Power	Station	Report Sent	Received	Signal Sent	Received

Remarks

Date	Time UTC Start	Finish	Frequency (MHz)	Mode	Power	Station	Report Sent	Received	Signal Sent	Received

Remarks

Date	Time UTC Start	Finish	Frequency (MHz)	Mode	Power	Station	Report Sent	Received	Signal Sent	Received

Remarks

AMATEUR RADIO STATION LOG

Date	Time UTC Start	Finish	Frequency (MHz)	Mode	Power	Station	Report Sent	Received	Signal Sent	Received

Remarks

Date	Time UTC Start	Finish	Frequency (MHz)	Mode	Power	Station	Report Sent	Received	Signal Sent	Received

Remarks

Date	Time UTC Start	Finish	Frequency (MHz)	Mode	Power	Station	Report Sent	Received	Signal Sent	Received

Remarks

Date	Time UTC Start	Finish	Frequency (MHz)	Mode	Power	Station	Report Sent	Received	Signal Sent	Received

Remarks

Date	Time UTC Start	Finish	Frequency (MHz)	Mode	Power	Station	Report Sent	Received	Signal Sent	Received

Remarks

Date	Time UTC Start	Finish	Frequency (MHz)	Mode	Power	Station	Report Sent	Received	Signal Sent	Received

Remarks

Date	Time UTC Start	Finish	Frequency (MHz)	Mode	Power	Station	Report Sent	Received	Signal Sent	Received

Remarks

Amateur Radio Station Log

Date	Time UTC Start	Finish	Frequency (MHz)	Mode	Power	Station	Report Sent	Received	Signal Sent	Received
Remarks										

Date	Time UTC Start	Finish	Frequency (MHz)	Mode	Power	Station	Report Sent	Received	Signal Sent	Received
Remarks										

Date	Time UTC Start	Finish	Frequency (MHz)	Mode	Power	Station	Report Sent	Received	Signal Sent	Received
Remarks										

Date	Time UTC Start	Finish	Frequency (MHz)	Mode	Power	Station	Report Sent	Received	Signal Sent	Received
Remarks										

Date	Time UTC Start	Finish	Frequency (MHz)	Mode	Power	Station	Report Sent	Received	Signal Sent	Received
Remarks										

Date	Time UTC Start	Finish	Frequency (MHz)	Mode	Power	Station	Report Sent	Received	Signal Sent	Received
Remarks										

Date	Time UTC Start	Finish	Frequency (MHz)	Mode	Power	Station	Report Sent	Received	Signal Sent	Received
Remarks										

AMATEUR RADIO STATION LOG

Date	Time UTC Start	Finish	Frequency (MHz)	Mode	Power	Station	Report Sent	Received	Signal Sent	Received
Remarks										

Date	Time UTC Start	Finish	Frequency (MHz)	Mode	Power	Station	Report Sent	Received	Signal Sent	Received
Remarks										

Date	Time UTC Start	Finish	Frequency (MHz)	Mode	Power	Station	Report Sent	Received	Signal Sent	Received
Remarks										

Date	Time UTC Start	Finish	Frequency (MHz)	Mode	Power	Station	Report Sent	Received	Signal Sent	Received
Remarks										

Date	Time UTC Start	Finish	Frequency (MHz)	Mode	Power	Station	Report Sent	Received	Signal Sent	Received
Remarks										

Date	Time UTC Start	Finish	Frequency (MHz)	Mode	Power	Station	Report Sent	Received	Signal Sent	Received
Remarks										

Date	Time UTC Start	Finish	Frequency (MHz)	Mode	Power	Station	Report Sent	Received	Signal Sent	Received
Remarks										

AMATEUR RADIO STATION LOG

Date	Time UTC Start	Finish	Frequency (MHz)	Mode	Power	Station	Report Sent	Received	Signal Sent	Received

Remarks

Date	Time UTC Start	Finish	Frequency (MHz)	Mode	Power	Station	Report Sent	Received	Signal Sent	Received

Remarks

Date	Time UTC Start	Finish	Frequency (MHz)	Mode	Power	Station	Report Sent	Received	Signal Sent	Received

Remarks

Date	Time UTC Start	Finish	Frequency (MHz)	Mode	Power	Station	Report Sent	Received	Signal Sent	Received

Remarks

Date	Time UTC Start	Finish	Frequency (MHz)	Mode	Power	Station	Report Sent	Received	Signal Sent	Received

Remarks

Date	Time UTC Start	Finish	Frequency (MHz)	Mode	Power	Station	Report Sent	Received	Signal Sent	Received

Remarks

Date	Time UTC Start	Finish	Frequency (MHz)	Mode	Power	Station	Report Sent	Received	Signal Sent	Received

Remarks

www.ingramcontent.com/pod-product-compliance
Lightning Source LLC
Chambersburg PA
CBHW080559030426
42336CB00019B/3250